SOFT X-RAY EMISSION FROM CLUSTERS OF GALAXIES AND RELATED PHENOMENA

ASTROPHYSICS AND SPACE SCIENCE LIBRARY

VOLUME 309

SOFT X-RAY EMISSION FROM CLUSTERS OF GALAXIES AND RELATED PHENOMENA

Edited by

RICHARD LIEU

University of Alabama,
Department of Physics,
Huntsville, U.S.A.

and

JONATHAN MITTAZ

University of Alabama,
Department of Physics,
Huntsville, U.S.A.

Springer-Science+Business Media, B.V.

A C.I.P. Catalogue record for this book is available from the Library of Congress.

ISBN 978-90-481-6681-7 ISBN 978-1-4020-2564-8 (eBook)
DOI 10.1007/978-1-4020-2564-8

Printed on acid-free paper

Contents

Contributing Authors

W.I.Axford	U Ala Huntsville	ian@axford.org
M.Bonamente	NASA NSSTC Huntsville	bonamem@email.uah.edu
R.Cen	Princeton U.	cen@astro.princeton.edu
S.Chakrabarti	Boston U	supc@bu.edu
S.Cola-Francesco	INAF Roma	coa@coma.mporzio.astro.it
D.De Young	NOAO Tuscon	deyoung@noao.edu
K.Dolag	U Padova	kdolag@pd.astro.it
F.Durret	IAP (Paris France)	durret@iap.fr
K.Dyer	NRAO Socorro	kdyer@nrao.edu
T.Fang	Carnegie Mellon U	fangt@cmu.edu
A.Finoguenov	MPE Graching	alexis@xray.mpe.mpg.de
R.Fusco-Femiano	IASF-CNR	dario@rm.iasf.cnr.it
M.Henriksen	U Maryland	henrikse@umbc.edu
J.S.Kaastra	SRON Utrecht	j.s.kaastra@sron.nl
R.Lieu	U Ala Huntsville	lieur@cspar.uah.edu
F.J.Lockman	NRAO Green Bank	jlockman@nrao.edu
D.Lumb	ESA ESTEC	dlumb@rssd.esa.int
D.McCammon	W Wiscinsin Madison	mccammon@wisp.physics.wisc.edu
J.P.D.Mittaz	U Ala Huntsville	mittazj@email.uah.edu
D.Nagai	U Chicago	daisuke@addjob.uchicago.edu
J.Nevalainen	Harvard CfA/SAO	jukka@head-cfa.harvard.edu
F.Nicastro	Harvard CfA/SAO	fnicastro@cfa.harvard.edu
R.Petre	NASA/GSFC	robert.Petre-1@nassa.gov
L.A.Phillips	Caltech	phillips@tapir.caltech.edu
W.Sanders	U Wisconsin Madison	sanders@physics.wisc.edu
B.Savage	U Wisconsin Madison	savage@astro.wisc.edu
K.Sembach	STScI	sembach@stsci.edu
M.Shull	U Colorado	mshull@spitzer.colorado.edu
J.Slavin	Harvard CfA/SAO	jslavin@cfa.harvard.edu
S.Snowden	NASA/GSFC	snowden@riva.gsfc.nasa.gov
Y.Takahashi	U Ala Huntsville	yoshi@cosmic.uah.edu
T.Tripp	Princeton U Observatory	tripp@astro.princeton.edu

Preface

Since the discovery in 1995 by the EUVE mission that some clusters of galaxies exhibit an EUV and soft X-ray emission component separate from the radiation of the hot virialized intracluster gas, the phenomenon was confirmed to be real by the ROSAT, BeppoSAX, and most recently the XMM-Newton missions. To date, approximately 30 % of the total sample of X-ray bright clusters located in directions of low Galactic HI absorption are found to have this soft excess behavior.

The theoretical development has principally revolved around two scenarios. First is the originally proposed thermal interpretation, which invokes large quantities of a warm ($T \sim 10^6$ K) gas - the amount is significant compared with the total baryonic matter budget in clusters. In current version of this model the warm gas is believed to be located primarily at the outer parts of a cluster where the density of the hot virialized gas is low. It may in fact correspond to that portion of the the Warm Hot Intergalactic Medium (WHIM) found in the vicinity of galaxy clusters - the nodes of high over density where filaments of warm intergalactic matter intersect.

In the second proposed scenario the cluster soft excess is interpreted as signature of a large population of cosmic rays, the electron component of which undergoes inverse Compton scattering with the cosmic microwave background to produce EUV and soft X-ray photons. The motive underlying behind this approach is to avoid enlisting too much extra baryonic matter, although it turns out that the cosmic rays introduce a pressure which is directly proportional to the measured EUV brightness of clusters, and which in many cases can be large enough to reach equipartition levels between relativistic particles and the thermal hot gas. Our present understanding of the role played by cosmic rays in the soft excess phenomenon is that at the centers of some clusters where the excess is extremely bright the cause is likely to be non-thermal, because the amount of thermal warm gas required to account for this excess is unacceptably large, and in any case the necessary emission lines that bear witness to their existence are absent. It is entirely plausible that the pressure of this central cosmic ray component, which primarily resides with the protons, is responsible for the discontinuation of the cooling flow, while the secondary electrons are the cause of the soft excess emission as they scatter off the microwave background.

This is the first meeting specifically devoted to the topic of cluster soft excess and related phenomena. It calls together a group of some 40 scientists, mostly experts in the field, to present their findings on the observational aspects of emission from clusters, signature of the WHIM in the cluster, intergalactic, and local environment, theoretical modeling of the WHIM in cosmological hydrodynamic simulations, theory and observations of cluster cosmic rays, magnetic fields, radio data and the use of the Sunyaev-Zeldovich effect as a means of constraining the important parameters. We are particularly grateful to F.J. Lockman and S.L. Snowden for presenting their latest picture of the cold gas distribution in the interstellar medium, as this affects our ability to model extragalactic EUV and soft X-ray data via Galactic absorption by HI and HeI. Papers given on future hardware concepts to further the spatial and spectral diagnosis of diffuse soft X-ray emission in general are also included.

Finally we express our thanks to Dana Ransom and Clay Durret at the University Relations of UAH for their design (both hardcopy and world-wide-web versions) of the conference logo and on-line registration facilities, to the UAH Physics Department secretaries Carolyn Schneider and Cindi Brasher for their help in printing the conference program and arranging the mini reception, and to the UAH computer support staff Joyce Looger and Wayne Tanev for their sterling effort in providing computer services to the delegates during the meeting. Lastly, Richard Lieu is indebted to his wife Anna for looking after the general well being of participants throughout this event.

RICHARD LIEU & JONATHAN MITTAZ

I

THE CLUSTER SOFT EXCESS

THE EXTREME ULTRAVIOLET EXCESS EMISSION IN FIVE CLUSTERS OF GALAXIES REVISITED

Florence Durret[1], Eric Slezak[2], Richard Lieu[3], Sergio Dos Santos[4], Massimiliano Bonamente[3]

[1]*Institut d'Astrophysique de Paris, CNRS, 98bis Bd Arago, 75014 Paris, France*

[2]*Observatoire de la Côte d'Azur, B.P. 4229, F-06304 Nice Cedex 4, France*

[3]*Department of Physics, University of Alabama, Huntsville AL 35899, USA*

[4]*SAp CEA, L'Orme des Merisiers, Bât 709, F-91191 Gif sur Yvette Cedex, France*

Abstract

Evidence for excess extreme ultraviolet (EUV) emission over a tail of thermal X-ray bremsstrahlung emission has been building up recently. In order to improve the signal to noise ratio in the EUV, we have performed a wavelet based image analysis for five clusters of galaxies observed both at EUV and X-ray energies with the EUVE and ROSAT satellites respectively. The profiles of the EUV and X-ray reconstructed images all differ at a very large confidence level and an EUV excess over a thermal bremsstrahlung tail is detected in all five clusters (Abell 1795, Abell 2199, Abell 4059, Coma and Virgo) up to large radii. Our results tend to suggest that the EUV excess is probably non thermal in origin.

1. Introduction

The Extreme Ultraviolet Explorer (EUVE) has detected emission from a few clusters of galaxies in the ~70-200 eV energy range. By order of discovery, these were: Virgo (Lieu et al. 1996b, Berghöfer et al. 2000a), Coma (Lieu et al. 1996a), Abell 1795 (Mittaz et al. 1998), Abell 2199 (Bowyer et al. 1998), Abell 4059 (Berghöfer et al. 2000b) and Fornax (Bowyer et al. 2001).

Excess EUV emission relative to the extrapolation of the X-ray emission to the EUV energy range was then detected in several clusters, suggesting that thermal bremsstrahlung from the hot ($\sim 10^8$K) gas responsible for the X-ray emission could not account entirely for the EUV emission. This excess EUV emission can be interpreted as due to two different mechanisms: thermal radiation from a warm ($10^5 - 10^6$K) gas, as first suggested by Lieu et al. (1996a), or inverse Compton emission of relativistic electrons or cosmic rays either on

R. Lieu and J. Mittaz (eds.), Soft X-Ray Emission from Clusters of Galaxies and Related Phenomena, 3–10.

Table 1. Exposure times and main cluster characteristics.

Cluster	EUVE total exp. time (s)	ROSAT exp. time (s)	Redshift	kT_X (keV)	Scaling (kpc/superpx)
A 1795	158689	33921	0.063	5.9	32.7
A 2199	93721	34633	0.0299	4.1	15.8
A 4059	145389	5225	0.0478	4.0	25.0
Coma	60822	20112	0.023	8.7	12.2
Virgo	146204	9135	0.003	2.4	1.6

the cosmic microwave background or on stellar light originating in galaxies (Hwang 1997; Bowyer & Berghöfer 1998; Ensslin & Biermann 1998; Sarazin & Lieu 1998; Ensslin et al. 1999), or both. The main difficulty with the thermal model is that since gas at typical cluster densities and in the temperature range $10^5 - 10^6$K cools very rapidly, a source of heating is necessary. Besides, recent results obtained with FUSE and XMM-Newton (see various contributions in these Proceedings) tend to indicate smaller quantities of warm gas and much smaller cooling flow rates than predicted from simulations. On the other hand, non thermal models (Bowyer & Berghöfer 1998, Sarazin & Lieu 1998, Ensslin et al. 1999, Lieu et al. 1999, Atoyan & Völk 2000, Brunetti et al. 2001) seem able to account for the excess EUV emission, but require very high cosmic ray pressure which may exceed the hot gas pressure (e.g. Lieu et al. 1999, Bonamente, Lieu & Mittaz 2001).

We present here a new analysis of EUV and X-ray images for five clusters, based on their wavelet denoising. Our aims are: to confirm the detection of EUV emission as far as possible from the cluster center; to derive accurate EUV and X-ray profiles; to confirm the reality and radial distributions of the soft excesses over thermal bremsstrahlung in these clusters.

2. The data and image analysis

Five clusters were observed with the EUVE satellite: Abell 1795, Abell 2199, Abell 4059, Coma and Virgo, and their ROSAT PSPC images were retrieved from the archive (see details in Table 1).

In order to have comparable spatial resolutions in the EUV and X-rays, all the images were rebinned to a "superpixel" size of 0.3077 arcminutes (18.5 arcseconds). The linear scale per superpixel at the cluster distance, estimated with $H_0 = 50$ km s^{-1} Mpc^{-1} and q_0=0 is given in Col. 6 of Table 1.

We first estimated the background level in several rectangles at the extremities of the EUV images. For all the clusters, the rectangular shape of the EUVE images, and in some cases the fact that we had two rectangular images (see Fig. 1) was a problem for any spatial analysis involving an isotropic examination of the data at scales larger than a fraction of the shorter side. In

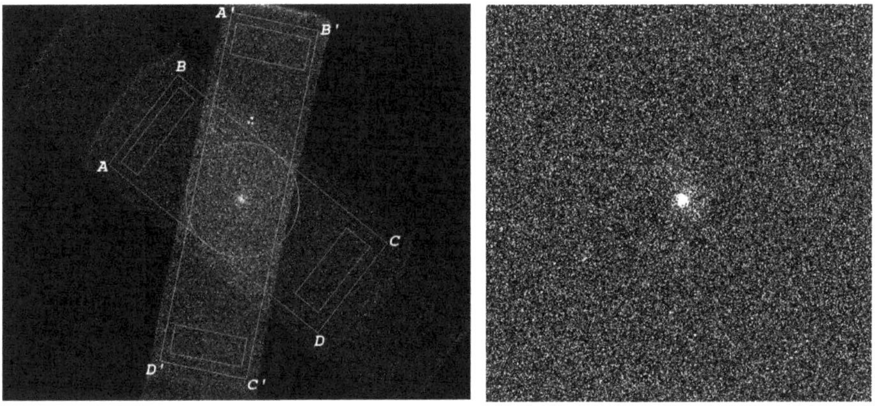

Figure 1. Abell 1795: left: initial EUVE image, right: "large" EUVE image (see text).

order to fill in the gaps and investigate without noticeable edge effects a scale as large as the shorter side of the EUVE rectangle, we created for each cluster a "large" square EUV image, keeping unchanged the central part of the image (a circle of $r \sim 16$ arcmin radius) and replacing each pixel outside this area by a random number drawn from a Poisson distribution, the variance of which is obtained by adding in quadrature the variances in each of the rectangles. The final image thus obtained for Abell 1795 is displayed in Fig. 1. Of course, profiles were only derived for radii smaller than the half width of the EUV detector, i.e. for $r < 16$ arcmin. Any features found within regions smaller than the PSFs (1 arcmin and 2.5 arcmin radii in the EUV and X-rays respectively) are of course unreliable.

The ROSAT data were reduced in the usual way using Snowden's software (Snowden et al. 1994). Since the geometry of these images is circular we had no problems here with the detector shape. The background was estimated as a mean value far from the edges of the detector and subtracted to the images before deriving the "raw" profiles (but not before making the wavelet analysis, since background subtraction is actually part of this process).

In order to increase the signal to noise ratio in the derived profiles, we applied part of the wavelet vision model described in detail in Rué & Bijaoui (1997), following these steps: first, a discrete wavelet transform of the image was performed up to about 1/3 the size of the square image; second, the noise was removed from the data; third a positive image where the background has been subtracted was restored. Profiles were then derived for the EUV and X-ray emissions, both on the raw (background subtracted) images and on the wavelet reconstructed ones, to check that the wavelet analysis and reconstruction did not modify the shape of the profile but only improved the signal to noise ratio. A point source (the star γTau) was analyzed in a similar way to see how the instrument point spread functions (PSF) could influence our results. The resulting PSFs drawn in the EUV and X-rays shown in Fig. 2 imply that no spatial information is reliable for radii smaller than $\sim$2.5 arcmin.

3. Results

X-ray and EUV profiles are displayed for the five clusters in Fig. 2. They show EUV emission as far as 16 arcmin radius (i.e. the central circle, see Section 2), comparable or larger than previously published values by other authors.

We performed a Kolmogorov-Smirnov test on the EUV and X-ray profiles drawn from the wavelet reconstructed images, after normalizing the EUV and X-ray profiles to the same innermost pixel value. In all clusters, the probability that the two profiles are issued from the same parent population is smaller than 0.1%, implying that statistically the EUV and X-ray profiles differ.

The observed EUV to X-ray intensity ratios (in the same concentric ellipses) are displayed for the five clusters in Fig. 3 together with the ratios predicted in the hot gas bremsstrahlung hypothesis. Note that for the three clusters in which XMM-Newton has shown the existence of a temperature gradient (Abell 1795, Coma and Virgo), these ratios remain virtually unchanged when this gradient is taken into account. All clusters show an EUV excess over hot thermal bremsstrahlung. This excess strongly increases with radius in Abell 1795, Abell 2199 and Abell 4059, and somewhat in Virgo. On the other hand, in Coma it remains roughly constant up to $r \sim 10$ arcmin, then decreases radially.

The ratios of the EUV to X-ray intensities for all clusters with radii expressed in physical units (kpc) are shown in Fig. 4. Note that the radial extent of the EUV and X-ray emission in Virgo is comparable to the other clusters when the radius is expressed in arcminutes, but becomes extremely small in physical distance units, due to the very small redshift of the cluster: slightly more than 80 kpc.

4. Discussion

We have shown unambigously the existence of a EUV excess in all five clusters of our sample. In the first three clusters (Abell 1795, Abell 2199 and Abell 4059), the EUV to X-ray intensity ratios show a possible deficiency of EUV emission over a bremsstrahlung tail in the very central regions (but this may be an effect of the ratio of the EUV to X-ray PSFs), and a clear EUV excess beyond a few arcmin. As suggested by several authors (Bowyer et al. 1999; Lieu et al. 2000), the first of these features, if real, can be interpreted as due to excess absorption within the cluster core due to the fact that in the cooler central regions some metals are not fully ionized and these ions absorb part of

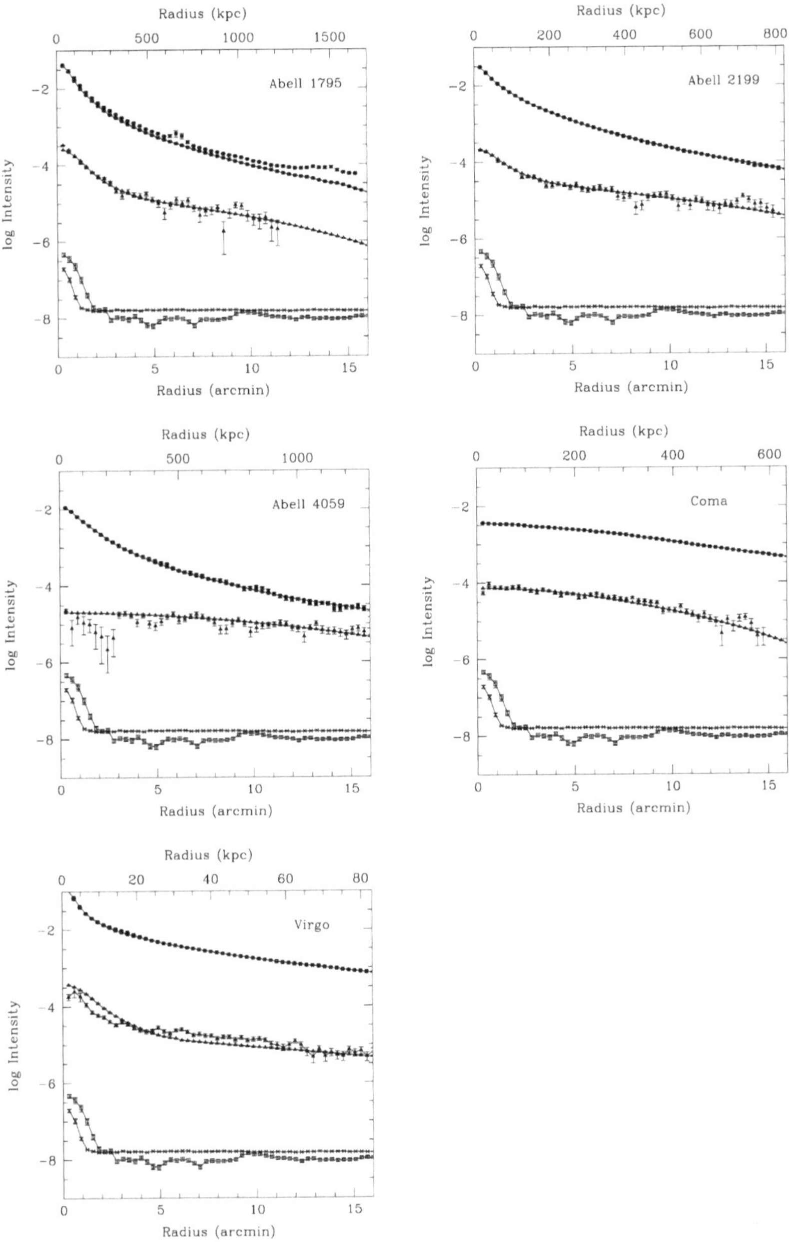

Figure 2. X-ray (top curves) and EUV (middle curves) profiles for all clusters, obtained from the raw background subtracted data (data with the largest rms error bars) and after a wavelet analysis followed by a reconstruction. The error bars for the profiles derived from the wavelet reconstructed images are drawn but they are too small to be clearly visible. The bottom lines show the EUV (x's) and ROSAT PSPC (empty squares) instrument PSFs.

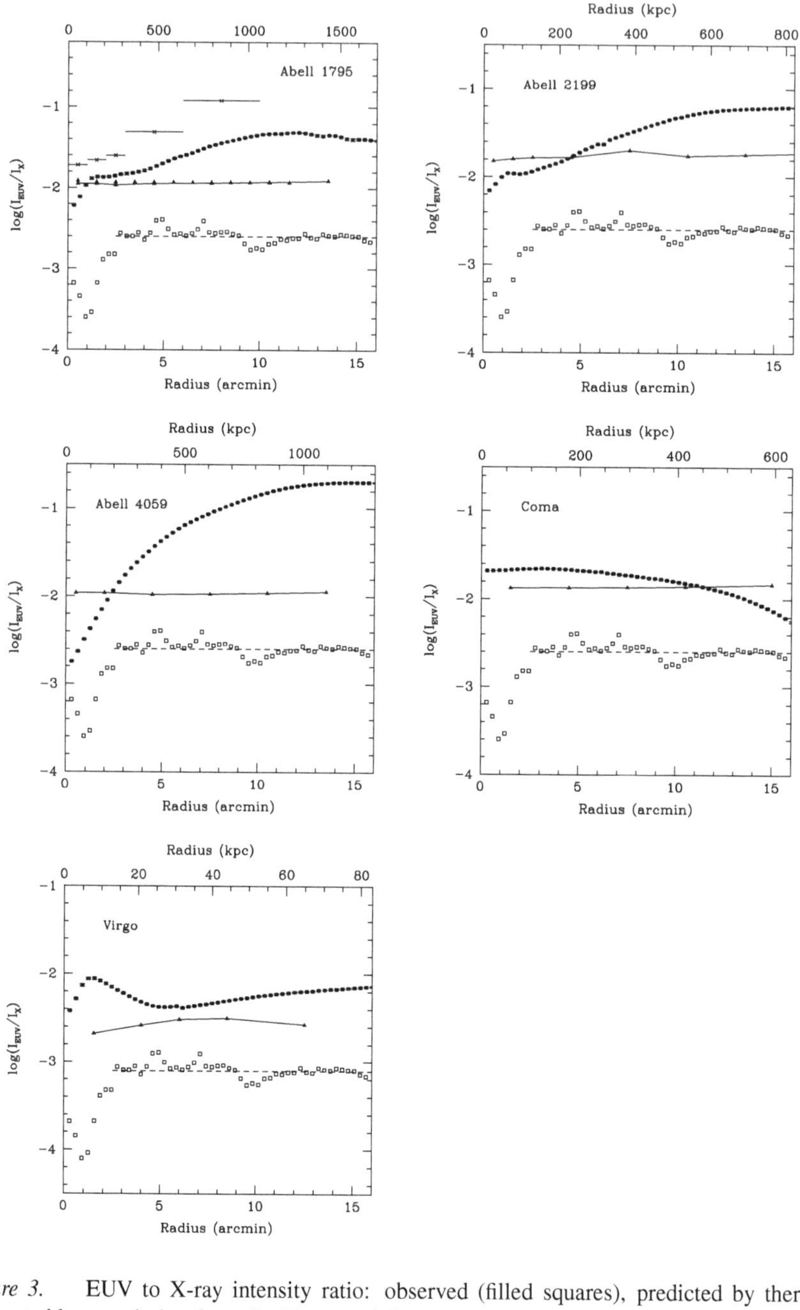

Figure 3. EUV to X-ray intensity ratio: observed (filled squares), predicted by thermal bremsstrahlung emission from the X-ray emitting gas (filled triangles), observed by Mittaz et al. (1998) in Abell 1795 (x's) and EUV to X-ray PSF ratio (empty squares) and mean value (dashed line).

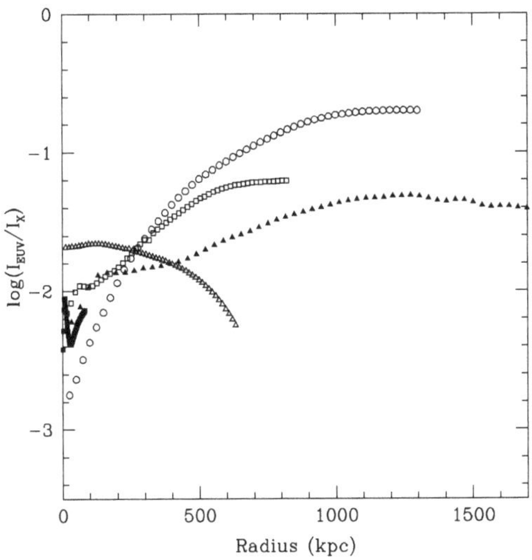

Figure 4. EUV to X-ray ratios for the five clusters in our sample, with the radius expressed in physical units (kpc). The symbols are the following: Abell 1795: filled triangles, Abell 2199: empty squares, Abell 4059: empty circles, Coma: empty triangles, Virgo: filled squares. Error bars are omitted for clarity.

the soft X-ray flux, in agreement with the existence of cooling flows in these three clusters. However, XMM-Newton has detected much weaker emission lines than expected from *bona fide* isobaric cooling flow models, implying that there is significantly less "cool" gas than predicted (see e.g. Kaastra et al. 2001 and these proceedings). Therefore it is no longer straightforward to interpret excess absorption at the center of these clusters as due to the presence of cooler gas.

The only cluster (Coma) for which no EUV dip is seen in the very center is both the hottest one by far and the only one with no "cooling flow" whatsoever. Since the ratio of the EUV to X-ray PSFs shows a dip for radii smaller than about 2.5 arcmin, the absence of such a dip in Coma suggests that in fact there is a significant EUV excess in the central regions of Coma.

Virgo shows a roughly constant EUV excess, somewhat stronger in the very center, suggesting as for Coma that the EUV excess is in fact strong in this zone. Images with higher spatial resolution are obviously necessary to analyze the EUV excess at these small radii.

The mere presence of an EUV excess in all five clusters of our sample indicates that a mechanism other than bremsstrahlung from the hot gas responsible for the X-ray emission is needed to account for this emission. Our data does not allow us to discriminate between the various mechanisms proposed in the literature to account for the EUV excess. However, in view of the most recent results obtained with XMM-Newton suggesting that there is much less warm gas in the central regions of clusters than previously believed (see in particular several presentations in these proceedings and references therein), it seems likely that this EUV excess is probably of non thermal origin, at least in the central regions.

A full description of the method and results presented here can be found in Durret et al. (2002).

Acknowledgments

We are very grateful to M. Traina for her invaluable help in obtaining the up-to-date version of the multiscale vision model package used throughout. We acknowledge discussions with D. Gerbal, G. Lima Neto, G. Mamon and J. Mittaz, and thank Jelle Kaastra for several interesting suggestions.

References

Atoyan A.M. & Völk H.J. 2000, ApJ 535, 45

Berghöfer T.W., Bowyer S. & Korpela E. 2000a, ApJ 535, 615

Berghöfer T.W., Bowyer S. & Korpela E. 2000b, ApJ 545, 695

Bonamente M., Lieu R. & Mittaz J. 2001, ApJ 561, L63

Bowyer S. & Berghöfer T.W. 1998, ApJ 506, 502

Bowyer S., Lieu R. & Mittaz J. 1998, in The Hot Universe, ed. K. Koyama, S. Kitamoto & M. Itoh (Dordrecht: Kluwer), 185

Bowyer S., Berghöfer T.W. & Korpela E.J. 1999, ApJ 526, 592

Bowyer S., Korpela E.J. & Berghöfer T.W. 2001, ApJ 548, 135

Brunetti G., Setti G., Feretti L. & Giovannini G. 2001, MNRAS 320, 365

Durret F., Slezak E., Lieu R., Dos Santos S., Bonamente M. 2002, A&A 390, 397

Ensslin T.A. & Biermann P.L. 1998, A&A 330, 90

Ensslin T.A., Lieu R. & Biermann P.L. 1999, A&A 344, 409

Hwang Z. 1997, Science 278, 1917

Kaastra J.S., Tamura T., Peterson J., Bleeker J. & Ferrigno C. 2001, Proc. Symposium "New visions of the X-ray Universe in the XMM-Newton and Chandra Era", 26-30 November 2001, ESTEC, The Netherlands

Lieu R., Mittaz J.P.D., Bowyer S. et al. 1996a, Science 274, 1335

Lieu R., Mittaz J.P.D., Bowyer S. et al. 1996b, ApJ 458, L5

Lieu R., Ip W.-H., Axford W.I. & Bonamente M. 1999, ApJ 510, L25

Lieu R., Bonamente M. & Mittaz J.P.D. 2000, A&A 364, 497

Mittaz J.P.D., Lieu R. & Lockman F.J. 1998, ApJ 498, L17

Rué F., Bijaoui A. 1997, Experimental Astronomy 7, 129

Sarazin C.L., Lieu R. 1998, ApJ 494, L177

Snowden S. L., McCammon D., Burrows D. N. & Mendehall J. A. 1994, ApJ 424, 714

SOFT (1 KEV) X-RAY EMISSION IN GALAXY CLUSTERS

Mark Henriksen
University of Maryland, Baltimore County

Abstract

In this paper, we present results from our program of analysis of the broad band spectra of galaxy clusters. For two clusters, Abell 754 and IC 1262, we find evidence of a soft excess that can be modeled either as thermal emission or as non-thermal emission. For IC 1262, the thermal interpretation is non-physical and the non-thermal model has further support from a radio halo that we propose exists in the cluster center. For Abell 754, we modeled the 1 keV thermal component as emission from the IGM of several galaxy groups that may lie within 300 kpc of the central region of the cluster. We further argue that this interpretation is more likely than either elliptical galaxies or intercluster gas as the origin of the emission. However, it is also possible that it is non-thermal in origin. Under the non-thermal interpretation, it is likely to originate from primary cosmic-rays accelerated at the shock front formed during the ongoing merger. Finally, we present the first observational L_{nt} versus Tx plot. The plot shows a weak (90%) correlation that indicates at least some of the non-thermal components detected in clusters must result from the formation of the cluster in order to provide this energy relationship.

1. Introduction: History of Non-thermal emission Detection

Over the first several decades of X-ray studies of galaxy clusters using broad band spectra, increasingly sensitive limits have been placed on the amount of non-thermal X-ray emission from these objects. Most of the early studies were restricted to clusters known to have diffuse radio halos since they have a population of non-thermal electrons. A wide range of data were used in this work (Rephaeli & Gruber 1988 using *HEAO1-A4*; Henriksen 1998 using *ASCA* and *HEAO 1-A2*; Rephaeli, Ulmer, & Gruber 1994 using OSSE on *CGRO*, and EGRET, Shreekumar et al. 1996). The upper limit on non-thermal X-ray emission together with the radio spectral parameters were used to place a lower limit on the average magnetic field, $<B>$. Lower limits to $<B>$ in the range of ≥ 0.1–0.2μG were found. The EGRET observation provided the highest lower

R. Lieu and J. Mittaz (eds.), Soft X-Ray Emission from Clusters of Galaxies and Related Phenomena, 11–19.

limit on the average cluster field (0.4μG) from these early observations with a flux upper limit of 4×10^{-8} ph cm^{-2} s^{-1} at >100 MeV. The next generation of hard X-ray observatories have given new and more sensitive limits on and several detections of non-thermal emission. For example, hard X-ray emission is detected above 25 keV with the *SAX* PDS identified with the flat spectrum, α_r = 0.8 radio source in A2256 (Fusco-Femiano et al. 2000).

Observations are needed at energies > 25 keV to detect a flat, non-thermal X-ray source in hot, ~7 keV clusters. Consequently, the *SAX* PDS has been the primary detector used to find hard X-ray components in A2256, A2199 (Kaastra et al. 1999), and Coma (Fusco-Femiano et al. 2000). In Abell 2199, the non-thermal component was also detected in the 8 - 10 keV range in the MECS (Kaastra et al. 1999). Non-thermal emission from cooler, less luminous clusters such as Abell 1750, Abell 1367, or IC 1262 can be detected with the RXTE PCA and the BeppoSaxMECS that provide high signal-to-noise up to 15 keV. For Abell 1367, the PCA is quite sensitive to non-thermal emission from the steep spectrum radio relic, α_r = 1.9, and gave a stringent upperlimit on non-thermal emission, 3.3×10^{-3} ph cm^{-2} keV^{-1} s^{-1} at 1 keV, provided that the hot gas component is 2 phase. The lowerlimit on the average magnetic field from the relic is very high, 0.84μG (Henriksen & Mushotzky 2001) and challenges the early notion that the average cluster magnetic field was in the range of 0.1 - 0.2; low enough to be created by galaxy wakes in the intracluster medium and points toward a more energetic process such as merger for magnetic field aplification. A low cluster temperature makes it possible to use the MECS to isolate and analyze small regions of the cluster and thereby minimize contamination from point sources while localizing the source of non-thermal emission. For a relatively steep spectrum source, as is typical of non-thermal radio halos, the non-thermal emission may also be detected as a soft excess over the cluster's thermal component. While the source flux of the powerlaw component is much higher in the low energy data and the sensitivity of the detectors is much superior to that at high energy, this strategy is complicated. In this case, the soft emission may represent the cooler phase of a two-phase intracluster medium. These competing hypotheses are explored in detail in the rest of this paper.

2. IC 1262: A Soft Component: Non-thermal Emission

We found preliminary indications of diffuse, non-thermal emission in IC 1262, a cool cluster (see Figure 1). using the BeppoSax MECS and ROSAT PSPC detectors (Hudson, Henriksen, & Colafrancesco 2003). By fitting a 6 arcmin (~360 h_{50}^{-1} kpc) region with a single mekal model including photoelectric absorption, we find a temperature of 2.1 - 2.3 keV, and abundance of 0.45 - 0.77 (both 90% confidence).

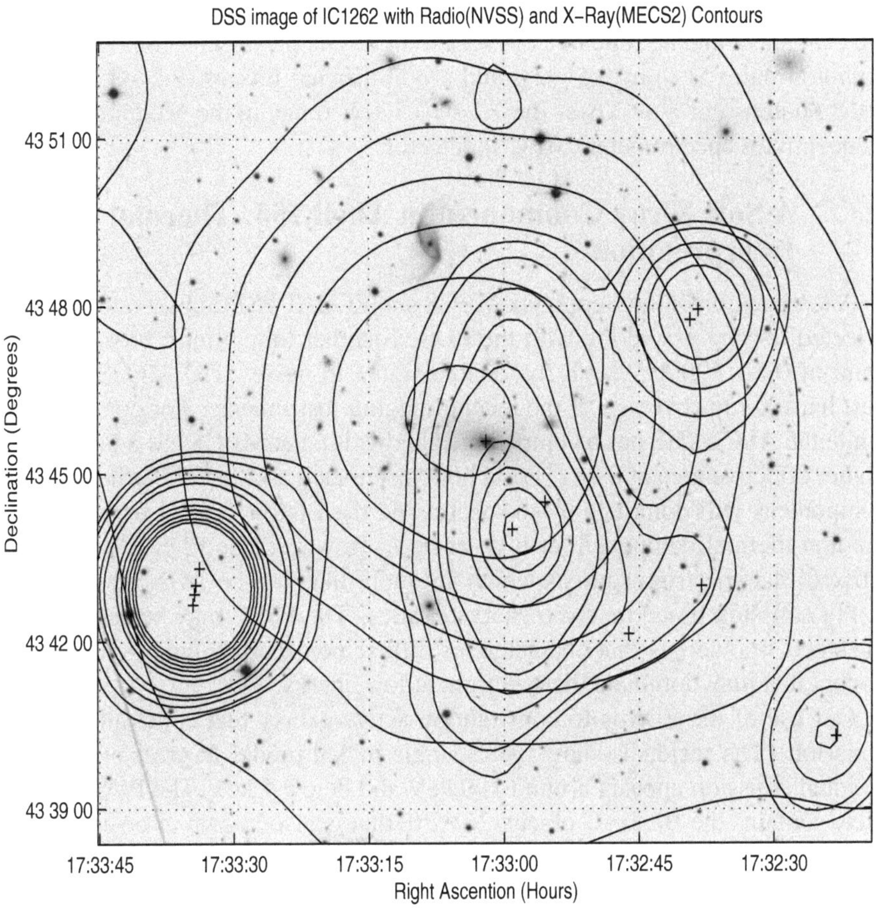

Figure 1. The North-South running contours near the center are radio intensity contours that delineate a radio halo.The other contours are X-ray and are the emission from the MECS.

The addition of a power-law component provides a statistically significant improvement (ftest = 90%)to the fit. The addition of a second thermal component also improves the fit though it is physically implausible based on the following argument. The two temperature components are such that the hottest component has a luminosity that is significantly lower than that predicted using the luminosity temperature relationship (Horner et al. 2000) to predict its temperature. Thus it is unphysical. Additional evidence of diffuse, non-thermal emission comes from the NRAO VLA Sky Survey (NVSS) and Westerbork Northern Sky Survey (WENSS) radio measurements, where excess diffuse, radio flux is observed after point source subtraction. The radio excess can be fit with a simple power-law, the spectral index of $\sim$1.8 which is consistent with the non-thermal X-ray emission spectral index. The steep spectrum is typical of diffuse emission and the size of the radio source implies that it larger than the cD galaxy and not due to a discreet source. The power-law component has a photon index (Γ_x) of 0.5 - 2.8 and a non-thermal flux of (2.2 - 7349.5) $\times$ 10^{-6} photons cm^{-2} s^{-1} over the 1.5 - 10.5 keV range in the Medium Energy Concentrator spectrometer (MECS) detector.

3. A Soft X-ray Component in Abell 754: Thermal Interpretation

A large region of the BeppoSax MECS and ROSAT PSPC observations were selected for analysis along with the PDS. Together they offer a broad energy band of 0.5 - 200 keV with reasonable signal to noise. The MECS has the best hard X-ray coverage of the recent imaging instruments. For comparison, while the ASCA GIS has a comparable bandwidth, the MECS has a factor of 2 higher effective area at 8 keV. For a hot cluster such as Abell 754, the thermal components will dominate these spectra and the PDS is crucial for detecting the non-thermal component at high energy. However, the PDS is most sensitive to flat spectrum sources which are more likely to be AGN than diffuse X-ray emission based on radio characteristics. The PSPC may be sensitive to the steeper powerlaw that characterizes diffuse non-thermal emission; a component that may dominate the spectrum at low energy.

Our use of the PDS is to constrain an active galaxy that contaminates the emission. The residuals show that a single mekal model fit gives significant residual emission appears around 100 keV and below 1 keV. The PDS field of view contains the BL LAC object, 26W20, that is also known to be a variable non-thermal X-ray source (Silverman, Harris, & Junor, 1998). After modeling 26W20 in the PDS, the residuals around 100 keV are eliminated. A better fit to the lower energy data is provided by adding either a non-thermal component or a second thermal component in addition to the high temperature thermal component. The second component is either a thermal component with temperature 0.75 - 1.03 keV and 7.0×10^{43}) ergs s^{-1} luminosity which is 7.2% of the the hot thermal component or non-thermal with $\alpha \simeq$2.3 and bolometric (0.1 - 100 keV) X-ray luminosity of 2.4×10^{42} or 0.25% of the thermal. There are a number of possible sources for cool thermal gas including: galaxies and intercluster gas.

Proposed sources of low temperature thermal emission of galactic origin include the integrated emission from gas associated with several elliptical galaxies or groups within the observed region of the cluster (Henriksen & Silk 1994). A number of small scale structures exist in the Abell 1367 cluster that are significantly cooler than their surroundings suggesting that the early-type galaxy coronae can survive in the intracluster medium (Sun & Murray 2002). We evaluated this hypothesis using the following simulations. The *Chandra* temperature map maintains approximately 6000 counts per region to allow spectral fitting of each region. The smallest regions are then 1 x 1 arcmin and the largest are 2 x 2 arcmin. At the redshift of Abell 754, 1 arcmin = 73 kpc (H_0 = 65). Recent XMM observations of NGC 5044, an elliptical galaxy in a group, will help to evaluate the importance of a single, large galaxy or group in providing the soft emission. These observations show that the 20 kpc region centered on the galaxy has temperature components of 0.7 and 1.1 keV, very similar to the cool thermal component in Abell 754. The luminosity from the entire group (r $<$ 250 kpc) is 1.8×10^{43} ergs s^{-1} (David et al (1994). Simulations (Henriksen & Hudson 2003) were performed to obtain an emission weighted temperature map for comparison with the Chandra map. The simulations consist of Abell 754 cluster emission with an embedded group whose gas is described by the NGC 5044 parameters (Mulchaey et al. 1996) The group is embedded at radii ranging from 100 to 400 kpc, which places it within the region of the BeppoSax data analyzed. The parameters for the group are kT = 0.98 keV, core radius = 28 kpc, ion density = 9.6e-3 cm^{-3}. The Abell 754 parameters are 1.85e-3 cm^{-3} (Abramopoulos & Ku 1981) and kT = 10 keV. The resulting simulated emission weighted temperature is (4.8, 5.1, 3.5, 2.5 keV) for a large pixel matched to the Chandra map and with a luminosity fraction of (24.6,22.8,37.5,50.4%) for the relative contributions of group versus cluster for a distance of (100,200,300,400 kpc). The emission weighted temperature at 400 kpc , 2.5 keV, is significantly less than 3.9 keV, the lowest temperature measured in the Chandra temperature. The group component begins to dominate the cluster component (luminosity fraction $>$ 50%) at this radius. This implies that the group(s) must be located within 300 kpc, otherwise the measured emission weighted temperature will contradict the Chandra temperature map constraint.

Because the luminosity of a single group is too low compared with the cool component we measured, we simulated the temperature map with 4 groups within 8' or a 580 kpc analyzed region. This would match the observed soft component luminosity. The groups must be further constrained to be within 300 kpc to avoid violating the constraint from the temperature map (e.g., the lowest measured temperature regions). The simulated temperature map gives one pixel slightly below 3.9 keV which does not violate the temperature map lower temperature limit. Thus we conclude that groups, like NGC 5044 may be the source of the cool component.

Elliptical galaxies have a hot gaseous halo of 1 keV. A King model using Abell 754 parameters (Abramopoulos & Ku 1983), with central surface density of galaxies of 92 Mpc^2 and the X-ray core radius for a King model (0.71 Mpc), gives 52 galaxies within the region. A typical elliptical fraction is 80%. This would give 42 elliptical galaxies. To give the cool component would require and average L_x of 1.7e-42 ergs s^-1 per elliptical galaxy. The distribution of normal early-type galaxies (Eskridge, Fabbiano, & Kim 1995) shows that 10^{41} ergs s^{-1} is typical and that only a small fraction have X-ray luminosities a factor of 10 higher. Thus it is implausible that this large cool component is from elliptical galaxies only.

Another possibility is a diffuse baryonic halo surrounding the cluster. Evidence for a similar, diffuse component in the 0.5 - 1 keV range have been reported with several different data sets and authors. Henriksen & White (1966) using HEAO-1 and Einstein data with a similar broad band coverage but with much lower effective area reported 0.5 - 1 keV gas beyond the central cooling flow region in several clusters and showed that the amount of the cool component did not correlate with the amount of cool central gas. There are several recent papers reporting cool emission within the cluster atmosphere.

Kaastra et al. (2003) report a soft X-ray component that is visible as an excess in the 0.4 - 0.5 keV range that is attributed to the warm hot IGM. Bonamente, Joy, and Lieu (2003) report evidence for a diffuse baryonic halo around the Coma cluster that extends to 2.8 Mpc. Their spectral analysis of ROSAT PSPC observations show that the radial temperature profile of this component is consistently 0.25 keV with 10% uncertainty. This suggests that the intercluster component is substantially cooler than the Abell 754 component since our temperature component is consistently higher. Nevalainen et al. (2003) report soft excesses in XMM observations of the Coma, Abell 2199, Abell 1795, and Abell 3112 in the cluster's inner 0.5 Mpc, a similar region to that analyzed in our BeppoSax and PSPC observations. The characteristic temperature is 0.6 - 1.3 keV, which is also consistent with the Abell 754 cool component. They point out that their density, 10^{-4} - 10^{-3} cm^{-3} is well above those expected from the intercluster medium (Dave 2001) in the core region of the cluster. The density of the Abell 754 cool component is similarly high compared to the WHIGM and comparable to the Coma, Abell 2199, and Abell 3122 clusters. Based on our analysis, we suggest that the soft 0.5 - 1 keV component in the central region of clusters, that is in excess of the projected intercluster component, is not due to the integrated emission from elliptical galaxy halos or a central cD galaxy but rather it is due to the integrated emission of the intergalactic medium of several embedded groups.Since most groups have a higher abundance than clusters, a test of this hypothesis would be to look for a high abundance in the soft component.

3.1 Soft X-ray Emission in Abell 754: Non-thermal Interpretation

The Abell 754 cluster shows evidence for a merger in its radio structure, X-ray surface brightness, and temperature map. The X-ray peak is surrounded by a peanut shaped, high surface brightness region. This region is cooler than the surrounding atmosphere by 30%, though it does not show a temperature discontinuity. Comparison of the morphology of X-ray center in the merger simulations (Roetigger, Stone, & Mushotzky 1998) to the that seen in the Chandra image shows a similarity in the steep density gradient to the EAST. The gradient is formed by ram-pressure as the subcluster hits the dense primary cluster core. However, the observed surface brightness also shows a steep gradient to the NW. This peanut shaped X-ray region isn't produced by the thermal gas motions as are shown in the merger simulations suggesting that another process contributes to giving that morphology. For example, the radio halo may be expanding and compressing the X-ray gas on the West side

The results of our analysis of the various X-ray data sets brings together new details of the dramatic cluster merger in Abell 754. Simulations of cluster merger (Takizawa & Naito 2000) trace the spatial distribution and time evolution of synchrotron radiation due to electrons accelerated at the shock fronts that form during a cluster merger. The morphology of the radio halo and its observed location relative to the shocked cluster gas depends on the viewing angle relative to the merger axis and the age of the merger. When the viewing perpendicular to the merger axis, the radio and thermal X-ray have different spatial relationships at different times in the merger. The shocked gas in Abell 754 West of the cluster center and the radio halo are co-spatial in the observations. That morphology seems to be consistent with the simulation morphology after maximum contraction. The simulations show two outward traveling shocks when the radio and shocked gas are co-spatial. While viewing nearly along the line of sight could also account for the shocked gas and radio being co-spatial, that viewing would also make both regions appear to be in the center of the cluster, which is not the case since they are West of center. Thus, we are likely viewing the merger perpendicular to the merger axis.

The Eastern radio source is not coincident with shocked gas and requires altering this merger scenario. Also, the X-ray peak region is not hot. However, this is consistent with the simulations of an off-center merger (Roettiger, Stone, & Mushotzky, 1998) that was hypothesized based on the ASCA temperature map (Henriksen & Markevitch 1996). A significant feature that is predicted and that we see inthe observations is a band of cool gas is produced running through the center. This gas is a mixture of pre-shocked primary cluster gas and cooler subcluster gas in the simulations. The Eastern radio source may be from an earlier merger or accretion event. Radio halos loose their energy due to inverse-Compton cooling and have a relatively short lifetime compared to the non-thermal hard X-ray component as shown in the simulations of Takizawa & Naito (2000). These simulations show that the radio halo fades quickly after the most contracting epoch, when the magnetic field is strongest, during the cluster merger. On the other hand, the 10 - 100 keV hard X-ray component reaches a peak after the radio when the cosmic-ray density peaks. Both occur only when there are signs of a merger present. Thus, it would be unlikely for the radio source to still be around after the thermal signs of the merger are gone. This would make the "previous merger" hypothesis unlikely. Ohno, Takizawa, & Shibata (2002) however find that turbulent Alfven waves may reaccelerate cosmic-rays that originated at the shock front thereby extending the lifetime of the radio halo. The turbulent gas mixing would also diminish the regions of hot gas from the shock front gas Secondary electrons produced by proton-proton collisions followed by pion decay may produce a radio halo with a longer lifetime (Blasi & Colafrancesco 1999). However, the spectral index is predicted to be flatter than we have observed from the X-ray. Thus, the present merger in Abell 754 and the radio structure, together with a hypothesized earlier merger to create the Eastern relic, make a non-thermal interpretation of the soft X-ray component physically plausible.

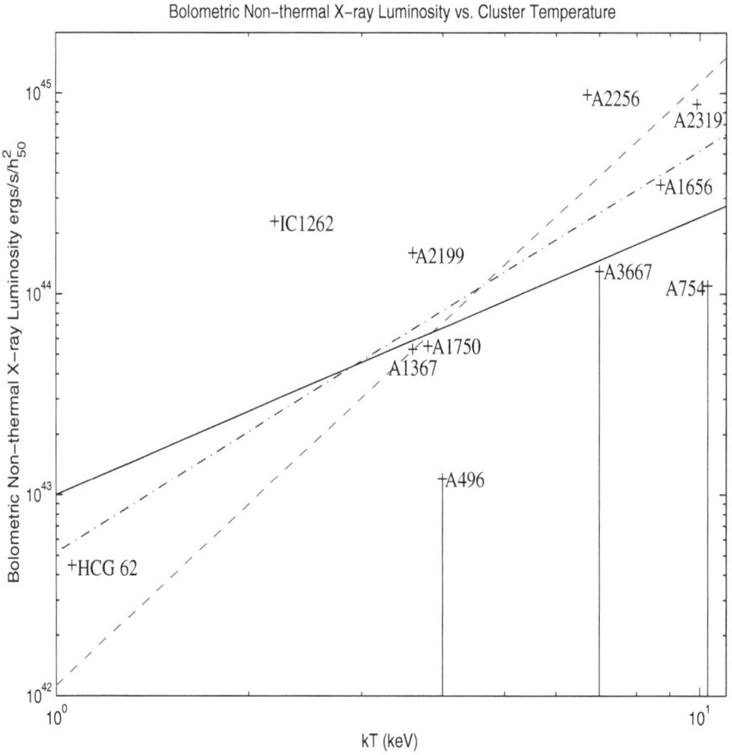

Figure 2. Clusters for which non-thermal X-ray measurements are available versus their emission weighted temperature.

In Figure 2, we show a plot of the non-thermal luminosity versus cluster temperature. The non-thermal X-ray luminosities represent all of the detections in the literatur. The figure shows that there is a correlation, as predicted by simulations of non-thermal emission by cosmic rays accelerated during cluster mergers (Miniati et al. 2001); but that there is a great deal of scatter. As this is a first generation plot of these two quantities, some of the scatter may be due to differences in analysis between authors. However, it does indicate that at least some of the detections must be real and not due to AGN, which would not show the correlation.

Acknowledgments
I thank Danny Hudson and Dr. Eric Tittley for contributions to this work.

References

Abramopoulos, F., & Ku, W.H.-M, 1983, ApJ, 271, 446.

Blasi, P., & Colafrancesco, 1999, APh, 12, 169.

Bonamente, M., Lieu, R., Joy, M., & Nevalainen, J., 2002, ApJ, 576, 688.

Brunetti, G., et al., 1999, in MPE Report 271: "Diffuse Thermal and relativistic Plasma in Galaxy Clusters", 263.

Dave, R., et al., 2001, ApJ, 552, 473.

David, L., Jones, C., Forman, W., Daines, S., 1994, ApJ, 428, 544.

De Grandi, S., Molendi, S., 2001, ApJ, 551, 153.

Eskridge, P. B.; Fabbiano, G., Kim, D.-W., 1995, ApJ Suppl., 97, 141.

Fusco-Femiano, R., et. al., 2002, astro-phg/0212408

Fusco-Femiano, R., et al., 2000, ApJ, 534, L7.

Giovannini, G., 1993, ApJ, 406, 399.

Henriksen, M., 1999, ApJ, 511, 666.

Henriksen, M., 1998, PASJ, 50, 389.

Henriksen, M., & Mushotzky, R., 2001, ApJ, 553, 84.

Henriksen & Markevitch, 1996, 1996, ApJ, 466, 79.

Henriksen, M., & Silk, J., 1996, in 'Clusters, Lensing, and the Future of the Universe', ASP Conference Series, 88, 213, Trimble & Reisenegger, eds.

Horner, D. J., Baumgartner, W. H., Gendreau, K.C., Mushotzky, R.F., Loewenstein, M., & Scharf, C.A. 2000, IAP 2000 meeting.

Kaastra, J., et al., 1999, ApJ, 519, 119.

Kaastra, J., Lieu, R., Tamura, T., Paerels, F., den Herder, J., 2003, Astron. & Astrophys., 397, 445

Miniati, F., Jones, T. W., Kang, H., Ryu, D., 2001, ApJ, 562, 233.

Mulchaey, J., Mushotzky, R., Burstein, D., Davis, D., 1996, ApJ, 456, 5.

Nevalainen, J., Lieu, R., Bonamente, M., & Lumb, D., 2003, ApJ, 584, 716.

Ohno, H., Takizawa, M., & Shibata, S., 2002, ApJ, 577, 685.

Rephaeli, Y., & Gruber, D., 1988, ApJ, 333, 133.

Rephaeli, Y., Ulmer, M., & Gruber, D., 1994, ApJ, 429, 554.

Roettiger, K., Stone, J., & Mushotzky, R., 1998, ApJ, 493, 62.

Silverman, J.D., Harris, D.E., & Junor, W. 1998, Astron. & Astrophys., 335, 443.

Sreekumar, P., et al., 1996, ApJ, 464, 628

Sun, M, & Murray, S., 2002, ApJ, 576, 708.

Takizawa, M., Naito, T., 2000, ApJ, 535, 586.

A MASSIVE HALO OF WARM BARYONS IN THE COMA CLUSTER

M. Bonamente[1], M.K. Joy[2] and R. Lieu[1]

[1]*Department of Physics, University of Alabama in Huntsville*

[2]*NASA/ MSFC Huntsville, Al.*

Abstract Several deep PSPC observations of the Coma cluster reveal a very large-scale halo of soft X-ray emission, substantially in excess of the well known radiation from the hot intra-cluster medium. The excess emission, previously reported in the central region of the cluster using lower-sensitivity EUVE and ROSAT data, is now evident out to a radius of 2.6 Mpc, demonstrating that the soft excess radiation from clusters is a phenomenon of cosmological significance. The X-ray spectrum at these large radii cannot be modeled non-thermally, but is consistent with the original scenario of thermal emission from warm gas at $\sim 10^6$ K. The mass of the warm gas is on par with that of the hot X-ray emitting plasma, and significantly more massive if the warm gas resides in low-density filamentary structures.

1. Introduction

In 1996, Lieu et al. (1996) reported the discovery of excess EUV and soft X-ray emission above the contribution from the hot ICM in the Coma cluster; their conclusions were based upon *Extreme UltraViolet Explorer (EUVE)* Deep Survey data (65-200 eV) and *ROSAT* PSPC data (0.15-0.3 keV). In this paper we present the analysis of a mosaic of *ROSAT* PSPC observations around the Coma cluster, revealing a very diffuse soft X-ray halo extending to considerably larger distances than reported in the previous studies. The spectral analysis of PSPC data reported in this paper indicates that the emission is very likely thermal in nature.

The redshift to the Coma cluster is z=0.023 (Struble and Rood 1999). Throughout this paper we assume a Hubble constant of $H_0 = 72$ km s^{-1} Mpc^{-1} (Freedman et al. 2001), and all quoted uncertainties are at the 68% confidence level.

R. Lieu and J. Mittaz (eds.), Soft X-Ray Emission from Clusters of Galaxies and Related Phenomena, 21–28.

2. The ROSAT PSPC data

Coma is the nearest rich galaxy cluster, and the X-ray emission from its hot ICM reaches an angular radius of at least 1 degree (e.g., White et al. 1993, Briel et al. 2001). Here we show how several off-center PSPC observations provide a reliable measurement of the background and reveal a very extended halo of soft X-ray excess radiation, covering a region several megaparsecs in extent in the Coma cluster.

The RASS maps of the diffuse X-ray background (Snowden et al. 1997) are suitable to measure the extent of the soft X-ray emission (R2 band, $\sim$ 0.15-0.3 keV) around the Coma cluster and to compare it with that of the higher-energy X-ray emission (R7 band, $\sim$ 1-2 keV). In Fig. 1 we show a radial profile of the surface brightness of PSPC bands R2 and R7 centered on Coma: the higher-energy emission has reached a constant background value at a radius of $\sim$ 1.5 degrees, while the soft X-ray emission persists out to a radius of $\sim$ 3 degrees. Thus, the RASS data indicate that the soft emission is more extended than the 1-2 keV emission, which originates primarily from the hot phase of the ICM.

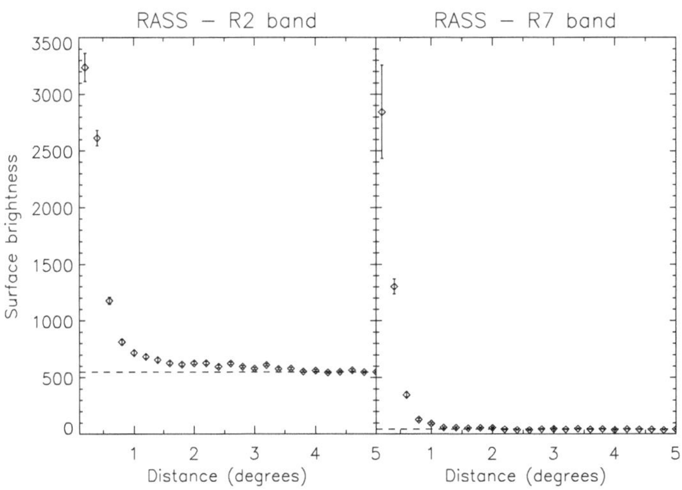

Figure 1. Radial profiles of soft X-ray emission (R2 band, 0.15-0.3 keV) and higher energy X-ray emission (R7 band, 1-2 keV) in the Coma cluster. Surface brightness units are 10^{-6} counts s^{-1} $arcmin^{-2}$ $pixel^{-1}$ (Snowden et al. 1998)

The all-sky survey data is based on very short exposures ($\sim$ 700 sec); therefore, for detailed studies of the soft X-ray emission in the Coma cluster, we use four pointed PSPC observations. Four additional deep PSPC observations are used to determine the local background. We show in section 3 below that the distribution of N_H is essentially constant within 5 degrees of the cluster center. Therefore, the off-source PSPC fields located 2.5-4 degrees away from the cluster center provide an accurate measurement of the soft X-ray background, and are also far enough away from the cluster center to avoid being contaminated by cluster X-ray emission (Fig. 1).

3. Galactic HI absorption in the direction of the Coma cluster

Knowledge of the Galactic absorption is essential to determine the intrinsic X-ray emission from extragalactic objects, particularly at energies $\leq$ 1 keV. Throughout this paper, we use the absorption cross sections of Morrison and McCammon (1983) in our models.

We use two methods to determine the distribution of neutral hydrogen in the Coma cluster. First, we use the radio measurements of Dickey and Lockman (1990) and of Hartmann and Burton (1997). The radio measurements are in excellent agreement, and show the measured HI column density varying smoothly from 9 $\times 10^{19}$ cm^{-2} to 11 $\times 10^{19}$ cm^{-2} within a radius of 5 degrees from the center of the Coma cluster. Second, we employ the far-infrared IRAS data, and use the correlation between 100 μm IRAS flux and HI column density (Boulanger and Perault 1988). The slope of the correlation is 1.2×10^{20} cm^{-2} / (MJy sr^{-1}), and the offset is determined by fixing the central N_H value to 9 $\times 10^{19}$ cm^{-2}, which is well established from independent radio measurements of the center of the Coma cluster (Dickey and Lockman 1990, Lieu et al. 1996, Hartmann and Burton 1997). The radial variation of N_H inferred from the IRAS data is in extremely good agreement with the radio measurements. The data indicate that
(a) the HI column density within the central 1.5 degree of the cluster is constant (9$\pm$1 $\times 10^{19}$ cm^{-2}), and
(b) in the region where the off-source background fields are located (2.5-4 degrees from the cluster center), the HI column density is between 9 $\times 10^{19}$ cm^{-2} and 11 $\times 10^{19}$ cm^{-2}. An N_H variation of this magnitude has a negligible effect on the soft X-ray flux in the PSPC R2 band (cf. Fig. 2 in Snowden et al. 1998).

4. Spectral analysis

The 4 PSPC Coma observations were divided into concentric annuli centered at R.A.=12h59'48", Dec.=25°57'0" (J2000), and the spectra were coadded to reduce the statistical errors. The pointed PSPC data were reduced according to the prescriptions of Snowden et al. (1994). For each of the 4 off-source fields in Fig. 1, a spectrum was extracted after removal of point sources. The spectra were statistically consistent with one another within at most $\sim$ 10 % point-to-point fluctuations. The off-source spectra were therefore coadded, and a 10% systematic uncertainty in the background was included in the error analysis. Further details on the data analysis can be found in Bonamente et al. (2002).

4.1 Single temperature fits

Initially, we fit the spectrum of each annulus in XSPEC, using a single-temperature MEKAL plasma model and the WABS Galactic absorption model. The results of the single temperature fit are given in Table 1 of Bonamente, Joy and Lieu (2003). If the neutral hydrogen column density is fixed at the Galactic value (9×10^{19} cm^{-2}, see section 3), the fits are statistically unacceptable (reduced χ^2 ranging from 3.1 to 9.7). Allowing the neutral hydrogen column density to vary results in an unrealistically low N_H for all of the annuli, and also produces statistically unacceptable fits (reduced χ^2 ranging from 2.5 to 4.2). We conclude that a single temperature plasma model does not adequately describe the spectral data, particularly at energies below 1 keV. Therefore, in the analysis that follows, we fit only the high energy portion of the spectrum (1-2 keV) with a single temperature plasma model, and introduce an additional model component to account for the low energy emission.

4.2 Modelling the hot ICM

To fit the high energy portion of the spectrum, we apply a MEKAL model to the data between 1 and 2 keV, and a photoelectric absorption model with $N_H = 9 \times 10^{19}$ cm^{-2}. The metal abundance is fixed at 0.25 solar for the central 20 arcmin region (Arnaud et al. 2001), and at 0.2 solar in the outer regions. The spectra are also subdivided into quadrants, in order to obtain a more accurate temperature for each region of the cluster. The results of the 'hot ICM' fit are given in Table 2 of Bonamente, Joy and Lieu (2003), and are consistent with the results previously derived from the PSPC data by Briel and Henry (1997), and with recent XMM measurements (Arnaud et al. 2001). In addition, the temperature found at large radii is in agreement with the composite cluster temperature profile of De Grandi and Molendi (2002).

4.3 Soft excess emission

The measured fluxes in the soft X-ray band can now be compared with the hot ICM model predictions in the 0.2-1 keV band. The results are shown in Fig. 2 and in Table 2 of Bonamente, Joy and Lieu (2003). The error bars reflect the uncertainty in the hot ICM temperature and the uncertainty in the Galactic HI column density ($N_H = 9 \pm 1 \times 10^{19}$ cm^{-2}). The soft excess component is detected with high statistical significance throughout the 90' radius of the pointed PSPC data, which corresponds to a radial distance of 2.6 Mpc. The soft excess emission (Fig. 2, left panel) is much more extended than that of the hot ICM (Fig. 2, right panel), in agreement with the conclusions drawn from the all-sky survey data (Fig. 1).

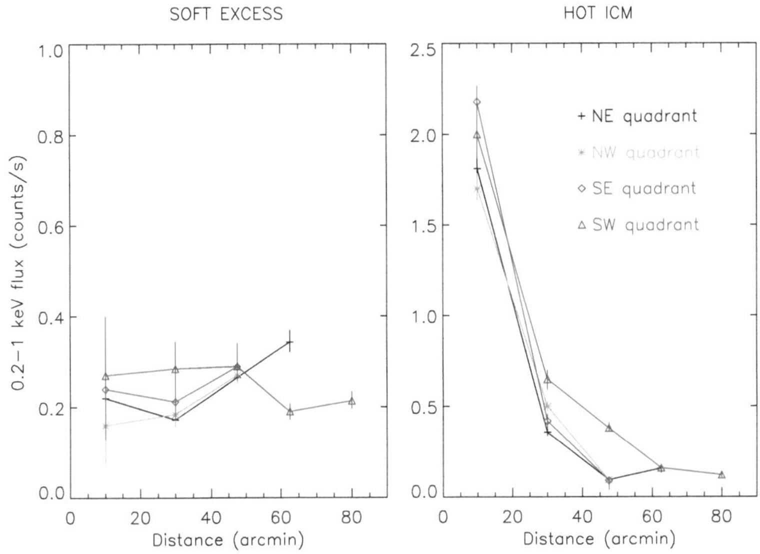

Figure 2. The radial distribution of the soft excess emission and of emission from the hot ICM

4.4 Low energy non-thermal component

Having established the hot ICM temperature for each quadrant, we now consider additional components in the spectral analysis. First, we add a power law non-thermal component, which predominantly contributes to the low energy region of the spectrum (Sarazin and Lieu 1998). The neutral hydrogen column density was fixed at the Galactic value ($N_H = 9 \times 10^{19}$ cm^{-2}). The results of fitting the hot ICM plus power law models to the annular regions are shown in Table 3 of Bonamente, Joy and Lieu (2003). The reduced χ^2 values are poor: the average χ^2_{red} is 1.48, and the worst case value is 1.89; we conclude that the combination of a low energy power law component and the hot ICM thermal model does not adequately describe the PSPC spectral data.

4.5 Low energy thermal component

Finally, we consider a model consisting of a 'hot ICM' thermal component (section 4.2) and an additional low-temperature thermal component. As before, the neutral hydrogen column density was fixed at the Galactic value (see section 3). The results of fitting the hot ICM plus warm thermal models are shown again in Table 3 of Bonamente, Joy and Lieu (2003). The reduced χ^2 values are significantly improved relative to the previous case: the average reduced χ^2 is 1.24 and the worst case value is 1.45. In every region the fit obtained with a warm thermal component was superior to the fit using a non-thermal component, as indicated by inspection of the χ^2_{red} values and by an F-test (Bevington 1969) on the two χ^2 distributions (Table 3 of Bonamente, Joy and Lieu 2003).

5. Interpretation

The spectral analysis of section 4 indicates that the excess emission can be explained as thermal radiation from diffuse warm gas. The non-thermal model appears viable only in a few quadrants, and will not be further considered in this paper.

5.1 A warm phase of the ICM

If the soft excess emission originates from a warm phase of the intra-cluster medium, the ratio of the emission integral of the hot ICM and the emission integral of the warm gas (Table 3 of Bonamente, Joy and Lieu 2003) can be used to measure the relative mass of the two phases. The emission integral is defined as

$$I = \int n^2 dV \, , \tag{1}$$

where n is the gas density and dV is the volume of emitting region (Sarazin 1988). The emission integral is readily measured by fitting the X-ray spectrum.

The emission integral of each quadrant determines the average density of the gas in that region, once the volume of the emitting region is specified (Eq. 1). Assuming that each quadrant corresponds to a sector of a spherical shell, the density in each sector can be calculated. The density of the warm gas ranges from $\sim 9\times10^{-4}$ cm^{-3} to $\sim 8 \times 10^{-5}$ cm^{-3}, and the density of the hot gas varies from 1.5×10^{-3} cm^{-3} to 6×10^{-5} cm^{-3}.

We assume that both the warm gas and the hot gas are distributed in spherical shells of constant density. Since the emission integral is proportional to $n^2 dV$ and the mass is proportional to ndV, the ratio of the warm-to-hot gas mass is

$$\frac{M_{warm}}{M_{hot}} = \frac{\int n_{warm} dV}{\int n_{hot} dV} = \frac{\int dI_{warm}/n_{warm}}{\int dI_{hot}/n_{hot}} \tag{2}$$

We evaluate Eq. 2 by summing the values of I_{hot}/n_{hot} and I_{warm}/n_{warm} for all regions (Tables 2 and 3 of Bonamente, Joy and Lieu 2003), and conclude that M_{warm}/M_{hot}=0.75 within a radius of 2.6 Mpc.

5.2 Warm filaments around the Coma cluster

It is also possible that the warm gas is distributed in extended low-density filaments rather then being concentrated near the cluster center like the hot ICM. Recent large-scale hydrodynamic simulations (e.g., Cen et al. 2001, Davé et al. 2001, Cen and Ostriker 1999) indicate that this is the case, and that 30-40 % of the present epoch's baryons reside in these filamentary structures. Typical filaments feature a temperature of T$\sim 10^5 - 10^7$ K, consistent with our results, and density of $\sim 10^{-5} - 10^{-4}$ cm^{-3} (overdensity of $\delta \sim 30 - 300$, Cen et al. 2001).

The ratio of mass in warm filaments to mass in the hot ICM is

$$\frac{M_{fil}}{M_{hot}} = \frac{\int n_{fil} dV_{fil}}{\int n_{hot} dV} = \frac{\int dI_{warm}/n_{fil}}{\int dI_{hot}/n_{hot}}. \qquad (3)$$

Assuming a filament density of $n_{fil} = 10^{-4}$ cm^{-3}, Eq. (3) yields the conclusion that M_{fil}/M_{hot}= 3 within a radius of 2.6 Mpc; the ratio will be even larger if the filaments are less dense. The warm gas is therefore more massive than the hot ICM if it is distributed in low-density filaments. More detailed mass estimates require precise knowledge of the filaments spatial distribution.

6. Conclusions

The analysis of deep PSPC data of the Coma cluster reveals a large-scale halo of soft excess radiation, considerably more extended than previously thought. The PSPC data indicate that the excess emission is due to warm gas at T$\sim 10^6$ K, which may exist either as a second phase of the intra-cluster medium, or in diffuse filaments outside the cluster. Evidence in favor of the latter scenario is provided by the fact that the spatial extent of the soft excess emission is significantly greater than that of the hot ICM.

The total mass of the Coma cluster within 14 Mpc is $1.6 \pm 0.4 \times 10^{15} M_\odot$ (Geller, Diaferio and Kurtz). The mass of the hot ICM is $\sim 4.3 \times 10^{14} M_\odot$ within 2.6 Mpc (Mohr, Mathiesen and Evrard 1999). The present detection of soft excess emission out to a distance of 2.6 Mpc from the cluster's center implies that the warm gas has a mass of at least $3 \times 10^{14} M_\odot$, or considerably larger if the gas is in very low density filaments. The PSPC data presented in this paper therefore lends observational support to the current theories of large-scale formation and evolution (e.g, Cen and Ostriker 1999), which predict that a large fraction of the current epoch's baryons are in a diffuse warm phase of the intergalactic medium.

References

Arnaud, M. et al. 2001, Astron. and Astrophys. , 365, L67
Bevington, P.R. 1969, Data reduction and error analysis for the physical sciences (McGraw-Hill)

Blandford, R.D. and Ostriker, J.P. 1978, Astron. and Astrophys. , 221, L29

Bonamente, M., Joy, M.K. and Lieu, R. 2003, Astrophys. Journal in press.

Bonamente, M., Lieu, R., Joy, M.K. and Nevalainen, J.H. 2002, Astrophys. Journal , 576, 688

Bonamente, M., Lieu, R. and Mittaz, J.P.D. 2001a, Astrophys. Journal Letters , 547,7

Boulanger, F. and Perault, M. 1988, Astrophys. Journal , 330, 964

Briel, U.G. et al. 2001, Astron. and Astrophys. , 365, L60

Briel, U.G. and Henry, J.P. 1997, *Proc. of the conference 'A New Vision of an old cluster: Untangling Coma Berenices' Eds. A. Mazure, F. Casoli, F. Durret and D. Gerbal*, pp. 170

Buote, D.A., 2001, Astrophys. Journal , 548, 652

Cen, R. and Ostriker, J.P. 1999, Astrophys. Journal Letters , 514, L1

Cen, R., Tripp, T.M., Ostriker, J.P. and Jenkins, E.B. 2001, Astrophys. Journal Letters , 559,L5

Davé, R., Cen, R., Ostriker, J.P., Bryan, G.L., Hernquist, L., Katz, N., Weinberg, D.H., Norman, M.L. and O'Shea, B. 2001, Astrophys. Journal , 552, 473

De Grandi, S. and Molendi, S. 2002, Astrophys. Journal , 567, 163

Dickey, J.M. and Lockman, F.J. 1990, Annual Revue of Astron. and Astrophys. , 28, 215

Freedman, W.L. et al. 2001, Astrophys. Journal , 553, 47

Geller, M.J., Diaferio, A. and Kurtz, M.J. 1999, Astrophys. Journal , 517, L26

Hartmann, D. and Burton, W.B. 1997, Atlas of Galactic Neutral Hydrogen (Cambridge: Cambridge Univ. Press)

Lieu, R., Ip, W.-I., Axford, W.I. and Bonamente, M. 1999b, Astrophys. Journal Letters , 510,L25

Lieu, R., Mittaz, J.P.D., Bowyer, S., Breen, J.O., Lockman, F.J., Murphy, E.M. & Hwang, C. -Y. 1996b, Science, 274,1335

Mohr, J.J., Mathiesen, B and Evrard, A. E. 1999, Astrophys. Journal , 517, 627

Morrison, R. and McCammon, D. 1983, Astrophys. Journal 270 119

Plucinsky, P.P, Snowden, S.L., Briel, U.G., Hasinger, G. and Pfefferman, E. 1993, Astrophys. Journal 418 519 Astron. and Astrophys. , 134, S287

Sarazin, C.L. 1988, X-ray emission from clusters of galaxies, Cambridge Astrophysics Series (Cambridge: Cambridge University Press)

Sarazin, C.L. and Lieu, R. 1998, Astrophys. Journal Letters , 494, L177

Snowden, S.L. et al. 1997, Astrophys. Journal , 485, 125

Snowden, S.L., Egger, R., Finkbeiner, D.P., Freyberg, M.J. and Plucinsky, P.P. 1998, Astrophys. Journal , 493, 715

Snowden, S.L., McCammon, D., Burrows, D.N. and Mendenhall, J.A. 1994, Astrophys. Journal 424 714

Struble, M.F. and Rood, H.J. 1999, Astrophys. Journal Supplements , 125, 35

Yan, M., Sadeghpour, H.R. and Dalgarno, A. 1998, Astrophys. Journal 496 1044

White, S.D.M., Briel, U.G. and Henry, J.P. 1993, Monthly Not. of the Royal Astron. Society , 261, L8

SOFT X-RAY EXCESS EMISSION IN THREE CLUSTERS OF GALAXIES OBSERVED WITH XMM-NEWTON

J. Nevalainen, [1,2,3] R. Lieu, [2], M. Bonamente, [2] and D. Lumb[3]

[1] *Harvard - Smithsonian Center for Astrophysics, Cambridge, USA,*

[2] *University of Alabama in Huntsville, Huntsville, USA,*

[3] *ESTEC, Noordwijk, Netherlands*

Abstract We present results on the spectroscopic analysis of XMM-Newton EPIC data of the central 0.5 h_{50}^{-1} Mpc regions of the clusters of galaxies Coma, A1795 and A3112. A significant warm emission component at a level above the systematic uncertainties is evident in the data and confirmed by ROSAT PSPC data for Coma and A1795. The non-thermal origin of the phenomenon cannot be ruled out at the current level of calibration accuracy, but the thermal model fits the data significantly better, with temperatures in the range of 0.6 – 1.3 keV and electron densities of the order of 10^{-4} – 10^{-3} cm^{-3}. In the outer parts of the clusters the properties of the warm component are marginally consistent with the results of recent cosmological simulations, which predict a large fraction of the current epoch's bayons located in a warm-hot intergalactic medium (WHIM). However, the derived densities are too high in the cluster cores, compared to WHIM simulations, and thus more theoretical work is needed to fully understand the origin of the observed soft X-ray excess.

Keywords: galaxies: clusters – X-rays: galaxies

INTRODUCTION

The soft X-ray excess emission has been reported in several clusters with *EUVE*, ROSAT and BeppoSAX instruments (e.g. Lieu et al. 1996a,b, 1999a,b; Bonamente et al. 2001a,b,c,d, 2002; Mittaz et al. (1998), Kaastra et al. 1999, 2002; Arabadjis and Bregman 1999; Buote 2001). The previously analyzed ROSAT PSPC data indicate that the excess emission is probably of thermal origin. The superior spectral resolution, the large bandpass coverage (0.2 – 10 keV) and the large collecting area of the XMM-Newton EPIC makes it well suited for studying the soft excess phenomenon. With EPIC, one can for the first time constrain simultaneously the properties of the hot gas and the soft component in clusters of galaxies. In this work we analyse the PN and

R. Lieu and J. Mittaz (eds.), Soft X-Ray Emission from Clusters of Galaxies and Related Phenomena, 29–36.

MOS data from the central 0.5 h_{50}^{-1} Mpc regions of clusters Coma, A1795 and A3112. We will present results on the detection and modeling of the soft component and outline a possible scenario responsible for this phenomenon.

1.1 DATA ANALYSIS

The data used in this work are taken from XMM-Newton public data archive (Coma and A1795) and from the Guaranteed Time program (A3112). We used the pipeline processed products of Coma and A1795 and processed the A3112 data using epchain and emchain tools in SAS 5.2.0. We filtered the flares using the $\geq$10 keV light curves and extracted the cluster spectra using only photons designated with patterns 0 for PN and 0 – 12 for MOS. Excising point sources and bad pixels, we obtained the spectra in two large radial bins, 0-0.2-0.5 h_{50}^{-1} Mpc For the energy redistibution of PN, MOS1 and MOS2 we use the calibration files epn_ff20_sY9.rmf , m1_r5_all_15.rmf and m2_r5_all_15.rmf , respectively. We created the effective area files with arfgen-1.44.4 tool within SAS distribution, using the calibration information available in March 2002.

For the background estimate, we used the blank sky data (Lumb et al. 2002) extracted at the same detector coordinates as the source spectra, using the same count rate criteria in the $>$ 10 keV band as for the data. We limit our spectral analysis to 0.3 – 7.0 keV band to ensure that at low energies the source signal level is more than 10 times above that of the background and 5 times at the highest energies. This is needed to avoid any errors in the background subtraction, as explained below. We use the 12 – 14 keV (PN) and 10 – 12 keV (MOS) band of source and background data to normalize the background to correspond the background level during the cluster observations.

1.2 PROBLEMS WITH ISOTHERMAL MODELING

We modeled the whole 0.3 - 7.0 keV energy band data of PN with a single MEKAL model (Mewe et al, 1995) absorbed by an HI column density (N_H), obtaining unacceptable fits (see Table 1 and Fig. 1). Addition of 5% systematic errors to the whole band leads to improvement of the fits, but yet still not to an acceptable level for all regions in the 3 clusters (see Table 1). Thus, at least one of the components involved in the above analysis, i.e. isothermal model or N_H from radio measurements may be wrong.

1.2.1 GALACTIC ABSORPTION

A possible source of the above residuals is application of the incorrect absorption to the isothermal emission model. To test this scenario, we allowed the N_H to vary as a free parameter and obtained statistically acceptable fits to PN data. However, the resulting N_H values (Table 1) are consistent with zero and significantly below the HI column densities measured by narrow beam ($20'$) radio observations at 21 cm wavelengths, where the typical uncertainty is $\sim$

Table 1. Results of single temperature MEKAL fit to 0.3 – 7.0 keV PN data using 0% or 5% systematic errors and various values of N_H The statistical errors of N_H is $\sim 1\times 10^{19}$ atoms cm^{-2}. The χ^2 and the 90% confidence interval of N_H are shown when N_H is treated as a free parameter

		0%	5%	5%, N_H free			
radii	d.o.f.	$\frac{\chi^2}{d.o.f.}$	$\frac{\chi^2}{d.o.f.}$	$\frac{\chi^2}{d.o.f.}$	$N_H{}^a$	$N_{HG}{}^b$	$N_{HG}{}^c$
			Coma				
0′– 5′	176	3.46	0.93	0.77	$0^{+0.1}_{....}$	0.9	0.9
5′– 13′	176	6.72	1.11	0.90	$0^{+0.1}_{....}$	0.9	0.9
			A1795				
0′– 2′	177	4.04	1.11	0.94	$0^{+0.2}_{....}$	1.0	1.2
2′– 5′	176	2.76	1.04	0.91	$0^{+0.3}_{....}$	1.0	1.2
			3112				
0′– 1.5′	177	3.32	1.48	1.06	$0.2^{+0.5}_{-0.2}$	2.5	
1.5′– 4.5′	177	1.90	1.28	1.04	$0.7^{+0.5}_{-0.6}$	2.5	

a: N_H as a free parameter, in 10^{20} atoms cm^{-2}
b: fine beam 21 cm N_H, in 10^{20} atoms cm^{-2} (Murphy et al. in prep.)
c: broad beam 21 cm N_H, in 10^{20} atoms cm^{-2} (Dickey & Lockman 1990))

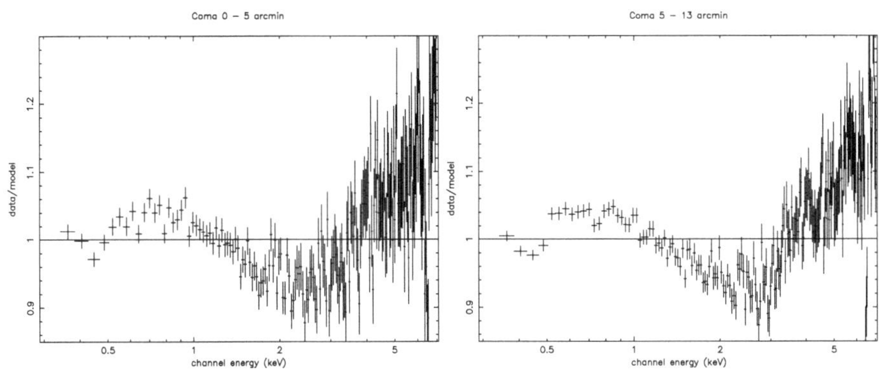

Figure 1. The ratio of the PN data of Coma to the best fit single temperature model to the 0.3 - 7.0 keV band

1×10^{19} cm^{-2} (Murphy et al. in prep.). To explain the residuals with incorrect N_H thus would render the radio data obsolete, which is not likely.

1.3 SOFT EXCESS

Extrapolating the 2 – 7 keV band best-fit isothermal models to soft X-ray energies reveals the soft excess (see Fig. 2) in Coma, A1795 and A3112. In all

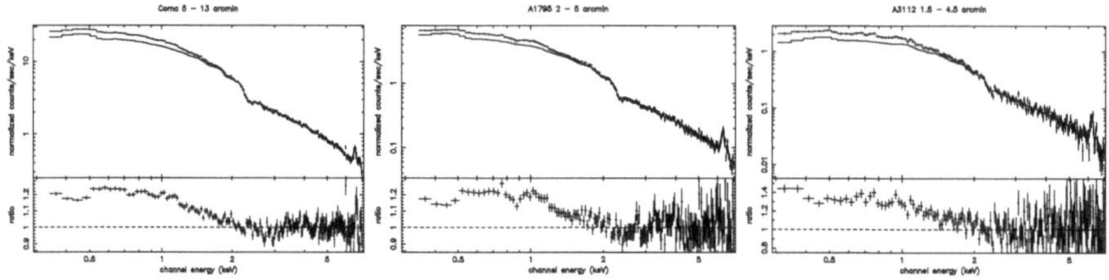

Figure 2. The 0.2 - 0.5 Mpc PN data of Coma, A1795 and A3112 with 1σ statistical uncertainties. The solid line shows the best fit single temperature fit to 2 – 7 keV data. Lower panels show the ratio of the data to the extrapolated model.

cases, both in PN and MOS, the data are above the model at energies below 2 keV. The excess increases towards lower energies, reaching 20% (40%) of the model level at 0.3 keV for Coma and A1795 (A3112). The difference of the soft excess fraction in A3112 is a further indication that the effect is not due to calibration problems.

Table 2. Properties of the thermal soft component. Luminosities are obtained in the 0.2 - 2.0 keV band and are in units of 10^{43} erg s^{-1}, using H = 50 km s^{-1} Mpc^{-1}.

	PN		MOS		PN + MOS			
radii	T [keV]	ab [solar]	T [keV]	ab [solar]	T [keV]	ab [solar]	L_{warm}	$\frac{L_{warm}}{L_{hot}}$
				Coma				
0′– 5′	$0.84^{+0.16}_{-0.10}$	$0.03^{+0.03}_{-0.02}$	$0.94^{+0.14}_{-0.12}$	$0.00^{+0.02}_{-...}$	$0.89^{+0.19}_{-0.15}$	$0.02^{+0.05}_{-0.02}$	$1.1^{+0.3}_{-0.2}$	$0.14^{+0.03}_{-0.03}$
5′– 13′	$0.82^{+0.13}_{-0.08}$	$0.03^{+0.03}_{-0.02}$	$0.90^{+0.11}_{-0.10}$	$0.00^{+0.02}_{-...}$	$0.86^{+0.15}_{-0.12}$	$0.02^{+0.04}_{-0.02}$	$3.1^{+0.6}_{-0.6}$	$0.15^{+0.03}_{-0.03}$
				A1795				
0′– 2′	$0.96^{+0.13}_{-0.10}$	$0.06^{+0.06}_{-0.03}$	$1.02^{+0.13}_{-0.14}$	$0.02^{+0.04}_{-0.02}$	$0.99^{+0.17}_{-0.12}$	$0.03^{+0.09}_{-0.03}$	$5.8^{+1.6}_{-1.8}$	$0.15^{+0.04}_{-0.05}$
2′– 5′	$0.86^{+0.18}_{-0.11}$	$0.03^{+0.04}_{-0.02}$	$0.88^{+0.17}_{-0.14}$	$0.00^{+0.01}_{-...}$	$0.87^{+0.18}_{-0.13}$	$0.06^{+0.01}_{-0.06}$	$3.6^{+1.2}_{-0.8}$	$0.14^{+0.05}_{-0.03}$
				A3112				
0′– 1.5′	$0.81^{+0.12}_{-0.07}$	$0.02^{+0.02}_{-0.02}$	$1.18^{+0.11}_{-0.14}$	$0.08^{+0.07}_{-0.06}$	$1.00^{+0.29}_{-0.26}$	$0.05^{+0.10}_{-0.05}$	$6.3^{+2.7}_{-1.7}$	$0.22^{+0.09}_{-0.06}$
1.5′– 4.5′	$0.85^{+0.12}_{-0.10}$	$0.02^{+0.02}_{-0.02}$	$0.75^{+0.10}_{-0.14}$	$0.00^{+0.02}_{-...}$	$0.80^{+0.17}_{-0.16}$	$0.01^{+0.03}_{-0.01}$	$3.4^{+0.9}_{-0.7}$	$0.22^{+0.06}_{-0.05}$

1.3.1 THERMAL MODELING

The PN and MOS data give consistent results for the temperature and metal abunbance at all cases. The results (Table 2, Fig. 3) indicate that the soft component has similar temperatures of 0.6 - 1.3 keV in different clusters inside 0.5 h_{50}^{-1} Mpc. The metal abundances are low, below 0.15 solar within uncertainties and in most cases consistent with zero. The luminosities of the warm

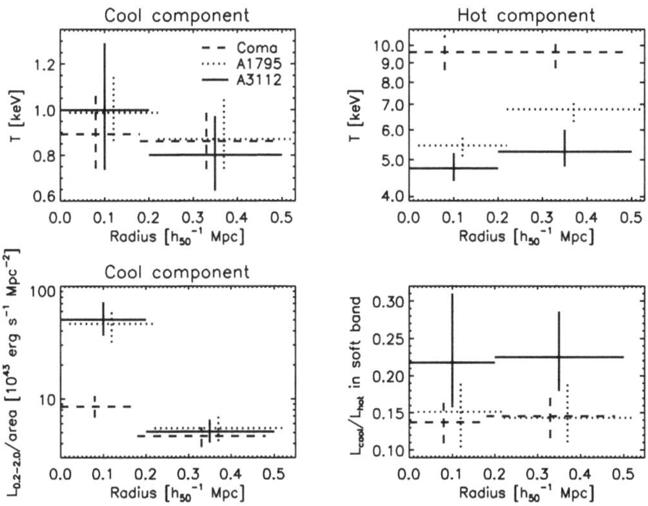

Figure 3. The best-fit values and 90% confidence uncertainties for the warm component using a thermal model (left panels). Upper right panel shows the temperatures of the hot component. Dashed, dotted and solid lines linescorrespond to Coma, A1795 and A3113, respectively. The radial bin values (0–0.2–0.5 Mpc) have been shifted slightly for display purposes. Lower right panel shows the ratio of the luminosities of the hot and warm component in 0.2 – 2.0 keV band.

component in 0.2 - 2.0 keV energy band are consistent in different clusters in radial range 0.2 - 0.5 h_{50}^{-1} Mpc. In the central 0.2 Mpc the 0.2 - 2.0 keV luminosities are consistent within the two cooling flow clusters A1795 and A3112, both being six times higher than in Coma. The 0.2 – 2.0 keV luminosities per metric area of the warm component increase by a factor of $\sim$ 10 between 0.2 – 0.5 and 0 - 0.2 Mpc in A1795 and A3112. Interestingly, the hot component behaves the same way, producing constant warm-to-hot component luminosity ratio in a given cluster in 0.2 - 2.0 keV energy band 0.5 h_{50}^{-1} Mpc, although a variation by a factor of 2 is allowed by the errors.

1.3.2 NON-THERMAL

Using a power-law model for the soft component leads to systematically poorer fits (see Table 3), but which are statistically acceptable in all cases, except for the center of A3112. Since most of the reduced χ^2 values for the thermal fits are below unity, this implies an overestimate of the systematic errors. If the calibration were better than the 5% level, and thus smaller systematic error would affect the model, the thermal fits would probably yield reduced χ^2 values of unity, and the power-law fits would all become unacceptable. Thus, although for the moment a firm conclusion of the nature of the soft excess is not available, the thermal model is preferred.

Table 3. Comparison of best-fits to PN data using mekal or power-law to model the soft excess

	mekal		power-law	
radii	$\frac{\chi^2}{d.o.f.}$	d.o.f.	$\frac{\chi^2}{d.o.f.}$	d.o.f.
		Coma		
$0' - 5'$	0.50	176	0.76	177
$5' - 13'$	0.56	176	0.87	177
		A1795		
$0' - 2'$	0.65	177	1.08	178
$2' - 5'$	0.73	176	0.96	177
		A3112		
$0' - 1.5'$	0.98	177	1.34	178
$1.5' - 4.5'$	0.87	177	1.13	178

1.4 SOFT COMPONENT INTERPRETATION

The resulting densities of the warm component are similar in different clusters at the same radii (2-4 $\times 10^{-3}$ atoms cm^{-3} at 0 - 0.2 h_{50}^{-1} Mpc and 0.6 - 0.7 $\times 10^{-3}$ atoms cm^{-3} at 0.2 - 0.5 h_{50}^{-1} Mpc), corresponding to overdensities of 1000 - 200 (in terms of the critical density). The cooling time, estimated using bremsstrahlung cooling function, is larger or comparable to the Hubble time. The pressure of the ideal (warm) gas, $\sim 10^{-12}$ erg cm^{-3} is an order of magnitude smaller than that of the hot gas.

1.5 CONCLUSIONS AND DISCUSSION

In galaxy clusters Coma, A1795 and A3112, we observe a significant soft X-ray emission component above the level of systematic uncertainties. The non-thermal origin of the soft excess cannot be ruled out at the current level of calibration accuracy, but the thermal model fits the data better. The warm gas is found to have temperatures of 0.6 – 1.3 keV inside 0.5 h_{50}^{-1} Mpc, consistent with the 90% distribution of the temperature values in Warm Hot Intergalactic Medium WHIM simulations (Dave et al., 2001). Within 0.2 and 0.5 h_{50}^{-1} Mpc, the derived electron densities ($\sim 10^{-4}$ cm^{-3}) are marginally consistent with the simulations, but in the cluster cores the derived densities are too high compared to those given by the simulations. This indicates that while WHIM may be a viable explanation for the soft excess at the outer regions of the clusters, it cannot explain the soft excess phenomenon entirely.

While the luminosities of the hot component in Coma, A1795 and A3112 vary substantially, those of the warm component outside the cool core region

Table 4. Physical properties of the thermal components. n is the atom density, δ is the gas density in terms of the critical density. P gives the gas pressure while t_{cool} gives the bremsstrahlung cooling time

	WARM				HOT			
radii	n^a	δ	t_{cool} (10^9 y)	P^b	n^a	δ	t_{cool} (10^9 y)	P^b
				Coma				
$0'-5'$	1.5	510	13	2.2	3.5	1180	18	55
$5'-13'$	0.6	210	30	0.9	1.4	460	46	21
				A1795				
$0'-2'$	3.2	970	6	5.2	7.4	2200	6	65
$2'-5'$	0.7	200	27	1.0	1.6	460	34	17
				A3112				
$0'-1.5'$	3.3	940	6	5.6	6.2	1760	7	48
$1.5'-4.5'$	0.7	190	27	0.9	1.2	330	40	9.6

a: [10^{-3} cm^{-3}]
b: [10^{-12} erg cm^{-3}]

are consistent in the three clusters being considered. The similarities suggest a common origin for the warm component, independent of the hot gas and its cooling. The derived values for the pressure of the hot component along the central 0.5 Mpc line of sight towards Coma, A1795 and A3112 are an order of magnitude higher than those of the warm gas, suggesting that they are not in contact. These requirements can be satisfied in a WHIM scenario where filaments do not penetrate the clusters, but rather form an external network. In this scenario, the density of the hot gas in clusters drops faster with radius than that of the WHIM filaments. Indeed, in Bonamente et al. (2002) the ROSAT PSPC data pointed to a radial increase of the soft excess. Thus, at ~ 1 h_{50}^{-1} Mpc the pressure equipartition may be attained and stable structures like filaments may be maintained. Outside the center, the homogenous distribution of filaments, projected in the cluster direction, produces constant luminosity per Mpc^2 in different clusters.

REFERENCES

Berghoefer,T.W. and Bowyer, S. 2002, ApJ, 565, L17
Bonamente, M., Lieu, R., Joy, M. & Nevalainen, J., 2002, ApJ, 576, 688
Bonamente, M., Lieu, R. & Mittaz, J., 2001a, ApJl, 547,7
Bonamente, M., Lieu, R. & Mittaz, J., 2001b, ApJL, 561, 63
Bonamente, M., Lieu, R., Nevalainen, J & Kaastra, J., 2001c, ApJL, 552, 7
Bonamente, M., Lieu, R. & Mittaz, J.,2001d, ApJ, 546, 805
Bowyer, S., Berghoefer, T.W. and Korpela, E.J. 1999, ApJ, 526, 592

Buote, D., 2001, ApJ, 548, 652
Cen, R. & Ostriker, J., 1999, ApJ, 514, 1
Cen, R., Tripp, T., Ostriker, J. & Jenkins, E., 2001, ApJL, 559, 5
Davé, R., Cen, R., Ostriker, J. et al., 2001, ApJ, 552, 473
De Grandi S., & Molendi, S., 2002, ApJ, 567, 163
Dickey, J. & Lockman, F., 1990, ARAA, 28, 215
Hwang, 1997, Sci 278, 1917
Kaastra, J., Lieu, R.,Bleeker, J., Mewe, R. and Colafrancesco, S. 2002, ApJ, 574, L1
Kaastra, J., Lieu, R., Mittaz, J., Bleeker, J., Mewe, R., Colafrancesco, S. and Lockman, F. 1999, ApJl, 519, L119
Lieu, R., Bonamente, M. and Mittaz, J.P.D. 1999a, ApJl, 517, L91
Lieu, R., Bonamente, M., Mittaz, J.P.D., Durret, F., Dos Santos, S. and Kaastra, J. 1999b, ApJl, 527,L77
Lieu, R., Mittaz, J.P.D., Bowyer, S., Lockman, F.J., Hwang, C. -Y., Schmitt, J.H.M.M. 1996a, ApJl, 458, L5
Lieu, R., Mittaz, J.P.D., Bowyer, S., Breen, J.O., Lockman, F.J., Murphy, E.M. & Hwang, C. -Y. 1996b, Science, 274,1335
Lieu, R., Axford, W. & Bonamente, M., 1999, ApJL, 510, 25
Lumb, D., Warwick, R., Page, M., De Luca, A., 2002, A&A, 389, 93L
Mittaz, J.P.D., Lieu, R. and Lockman, F.J. 1998, ApJl, 498, L17
Mewe R., Kaastra J. & Liedahl D., 1995, "Update on meka: mekal", Legacy 6, 16
Sarazin, C. & Lieu, R., 1998, ApJL, 494, 177

ACKNOWLEDGMENTS

J. Nevalainen acknowledges an ESA Research Fellowship, and a NASA grant NAG5-9945. M. Bonamente gratefully acknowledges NASA for support. We thank Drs. J. Kaastra and M. Markevitch for useful comments.

XMM-NEWTON DISCOVERY OF O VII EMISSION FROM WARM GAS IN CLUSTERS OF GALAXIES

Jelle S. Kaastra[1], R. Lieu[2], T. Tamura[1], F.B.S. Paerels[3] and J.W.A. den Herder[1]

[1]*SRON National Institute for Space Research, Utrecht, The Netherlands*

[2]*University of Alabama at Huntsville, USA*

[3]*Columbia University, New York, USA*

Abstract XMM-Newton recently discovered O VII line emission from ~ 2 million K gas near the outer parts of several clusters of galaxies. This emission is attributed to the Warm-Hot Intergalactic Medium. The original sample of clusters studied for this purpose has been extended and two more clusters with a soft X-ray excess have been found. We discuss the physical properties of the warm gas, in particular the density, spatial extent, abundances and temperature.

1. Introduction

Soft excess X-ray emission in clusters of galaxies was first discovered using EUVE DS and Rosat PSPC data in the Coma and Virgo cluster (Lieu et al. 1996a,b). It shows up at low energies as excess emission above what is expected to be emitted by the hot intracluster gas, and it is often most prominent in the outer parts of the cluster. However, a serious drawback with the old data (either EUVE or Rosat) concerns their spectral resolution, which does not exist for the EUVE DS detector, and is very limited for the Rosat PSPC at low energies where the width of the instrumental broadening causes a significant contamination of the count rate by harder photons.

The forementioned reasons render it very difficult for firm conclusions about the nature of the soft excess emission to be deduced from the original data alone. For example, already in the first papers it was suggested that the emission may have a thermal origin, but is also consistent with it having a power law spectrum caused by Inverse Compton scattering of the cosmic microwave background on cosmic ray electrons (Sarazin & Lieu 1998).

With the launch of XMM-Newton it is now possible to study the soft excess emission with high sensitivity and with much better spectral resolution using the EPIC camera's of this satellite. The high resolution Reflection Grating

R. Lieu and J. Mittaz (eds.), Soft X-Ray Emission from Clusters of Galaxies and Related Phenomena, 37–44.

Spectrometer (RGS) of XMM-Newton has proven to be extremely useful in studies of the central cooling flow region, but due to the very extended nature of the soft excess emission, the RGS is not well suited to study this phenomemon.

2. XMM-Newton observations

XMM-Newton has by now observed a large number of clusters. We investigated the presence of soft excess emission in a sample of 14 clusters of galaxies. This work has been published by Kaastra et al. (2003a). In that paper the details of the data analysis are given. Briefly, much effort was devoted to subtracting properly the time-variable soft proton background, as well as the diffuse X-ray background. We made a carefull assesement of the systematic uncertainties in the remaining background, since a proper background subtraction has been one of the contentious issues in the discussion around the discovery of the soft excess in EUVE and Rosat data. In a similar way, the systematic uncertainties in the effective area and instrumental response of the EPIC camera's were carefully assessed and quantified. In the spectral fitting procedures, both the systematic uncertainties in the backgrounds and the instrument calibration were taken into account. Spectra were accumulated in 9 concentric annuli between 0 and 15 arcmin from the center of the cluster.

The original sample used by Kaastra et al (2003a) has been extended to 21 clusters using archival data (see also Kaastra et al. 2003c). These additional clusters were analyzed in exactly the same way as the original 14 clusters. The spectra were initially analyzed using a two temperature model for the hot gas, with the second temperature of the component fixed at half that of the first component. From experience with cooling flow analysis (Peterson et al. 2003; Kaastra et al 2003b) we learnt that such a temperature parameterization is sufficient to characterise fully the cooling gas in the cores of clusters, while in the outer regions it is an effective method to take the effects of eventual non-azimuthal variations in the annular spectra into account. We note that the temperature of the coolest "hot gas" component in all cases where we detect a soft excess is much higher than the (effective) temperature of the soft excess. This is due to the now well-known fact that the emission measure distribution of the cooling flow drops off very rapidly. In fact, our models for the cooling flow predict no significant emission from O VII ions in the cooling plasma (at least below the detection limit of XMM-Newton).

The presence of a soft excess in this sample of clusters was tested by formally letting the Galactic absorption column density be a free parameter in the spectral fitting. Of the 21 clusters, 5 have apparent excess absorption. All these 5 clusters are located in regions where dust etc. is important, or they have a very compact core radius such that the temperature gradients in the core

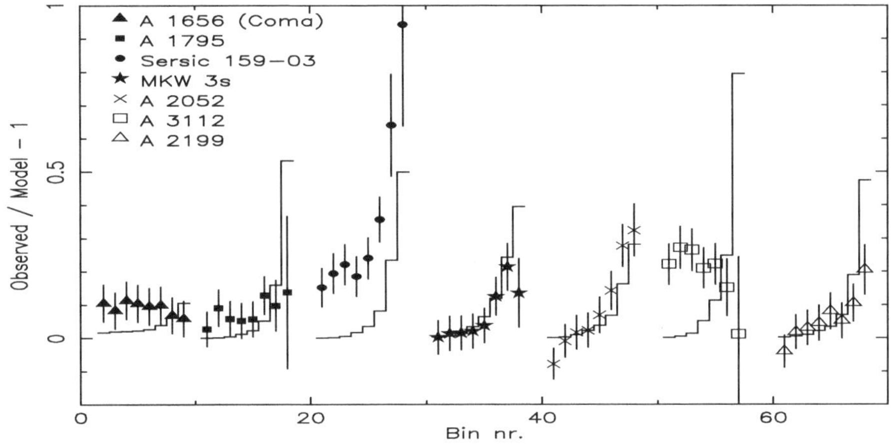

Figure 1. Soft excess in the 0.2–0.3 keV band as compared with a two temperature model (data points with error bars). The solid histogram is the predicted soft excess based upon scaling the sky-averaged XMM-Newton background near the cluster with the relative enhancement of the soft X-ray background derived from the 1/4 keV PSPC images. For the kth cluster the values for annulus j are plotted at bin number $10(k-1)+j$.

are not fully resolved by XMM-Newton and therefore the spectra are highly contaminated.

While the excess absorption in 5 clusters can be fully explained, the absorption deficit in 7 of these clusters cannot be explained by uncertainties in the calibration, background emission or foreground absorption, but only by the presence of an additional emission component. In fact, in several of the clusters the best-fit column density is zero!

In Fig. 1 we show the soft excess in the 0.2–0.3 keV band for the seven clusters with a significant soft component. These clusters are Coma, A 1795, Sérsic 159−03, MKW 3s, A 2052 (see also Kaastra et al. 2003a); A 3112 (see also Nevalainen et al. 2003 and Kaastra et al. 2003c) and A 2199 (paper in preparation).

The field of view of XMM-Newton is relatively small (∼15 arcmin radius) and therefore all these relatively nearby clusters fill the full field of view. For this reason, only a sky-averaged soft X-ray background (obtained from deep fields) as well as the time variable soft proton background (which is relatively small at low energies) were subtracted from the XMM-Newton data. Using the 1/4 keV Rosat PSPC sky survey data (Snowden et al. 1995), maps at 40 arcmin resolution were produced to estimate the average soft X-ray background in an annulus between 1–2 degrees from the cluster. Most of these seven clusters show an enhanced 1/4 keV count rate in this annulus (as compared to the typical sky-averaged 1/4 keV count rate). This, combined with the decreasing density of the hot gas in the outer parts of the cluster causes the apparent soft excess with increasing relative brightness at larger radii in Fig. 1. The figure shows complete consistency of the XMM-Newton data with the PSPC 1/4 keV data in this respect for A 1795, Sérsic 159−03, MKW 3s, A 2052 and A 2199. We show below that in these clusters this large-scale soft X-ray emission is due to thermal emission from the (super)cluster environment.

However, there is an additional soft component in Coma, A 3112, Sérsic 159−03 and perhaps A 1795. The fact that this component is above the prediction from the large scale PSPC structures implies that its spatial extent is at most 10–60 arcmin. We shall return to this component in Sect. 5.

3. Emission from the Warm-Hot Intergalactic Medium

In the previous section we found that five clusters show a soft excess in the 0.2–0.3 keV band at a spatial scale of at least 1–2 degrees, combining our XMM-Newton spectra with PSPC 1/4 keV imaging. It is not obvious a priori whether this excess emission is due to emission from the cluster region or whether it has a different origin, for example galactic foreground emission. Here the spectral resolution of the EPIC camera's is crucial in deciding which scenario is favoured. In Fig. 2 we show the fit residuals of the fit with two hot components and Galactic absorption only, in the outer 4–12 arcmin region combining all three EPIC camera's. The fit residuals show two distict features: a soft excess below 0.4–0.5 keV, and an emission line at ~0.56 keV. This emission line is identified as the O VII triplet, and detailed spectral fitting shows that both phenomena (soft excess and O VII line) can be explained completely by emission from a warm plasma with a temperature of 0.2 keV (see Kaastra et al. 2003a for more details). Thus, the soft excess has a thermal origin. Moreover, the centroid of the O VII triplet (which is unresolved) agrees better with an origin at the redshift of the cluster than with redshift zero. This clearly shows that the thermal emission has an origin in or near the cluster, although a partial contribution from Galactic foreground emission cannot be excluded in all cases. We also note that in A 1795 and A 2199 the O VII line is relatively weak and needs more confirmation.

Taking this additional soft thermal component into account in the spectral fitting yields fully acceptable fits. In fact, in the energy band below 1 keV, the soft component contributes 20–40 % of the X-ray flux of the outer (4–12 arcmin) part of the cluster!

We identify this component as emission from the Warm-Hot Intergalactic Medium (WHIM). Numerical models (for example Cen & Ostriker 1999, Fang et al. 2002) show that bright clusters of galaxies are connected by filaments that contain a significant fraction of all baryonic matter. Gas falls in towards the clusters along these filaments, and is shocked and heated during its accretion onto the cluster. Near the outer parts of the clusters the gas reaches its highest

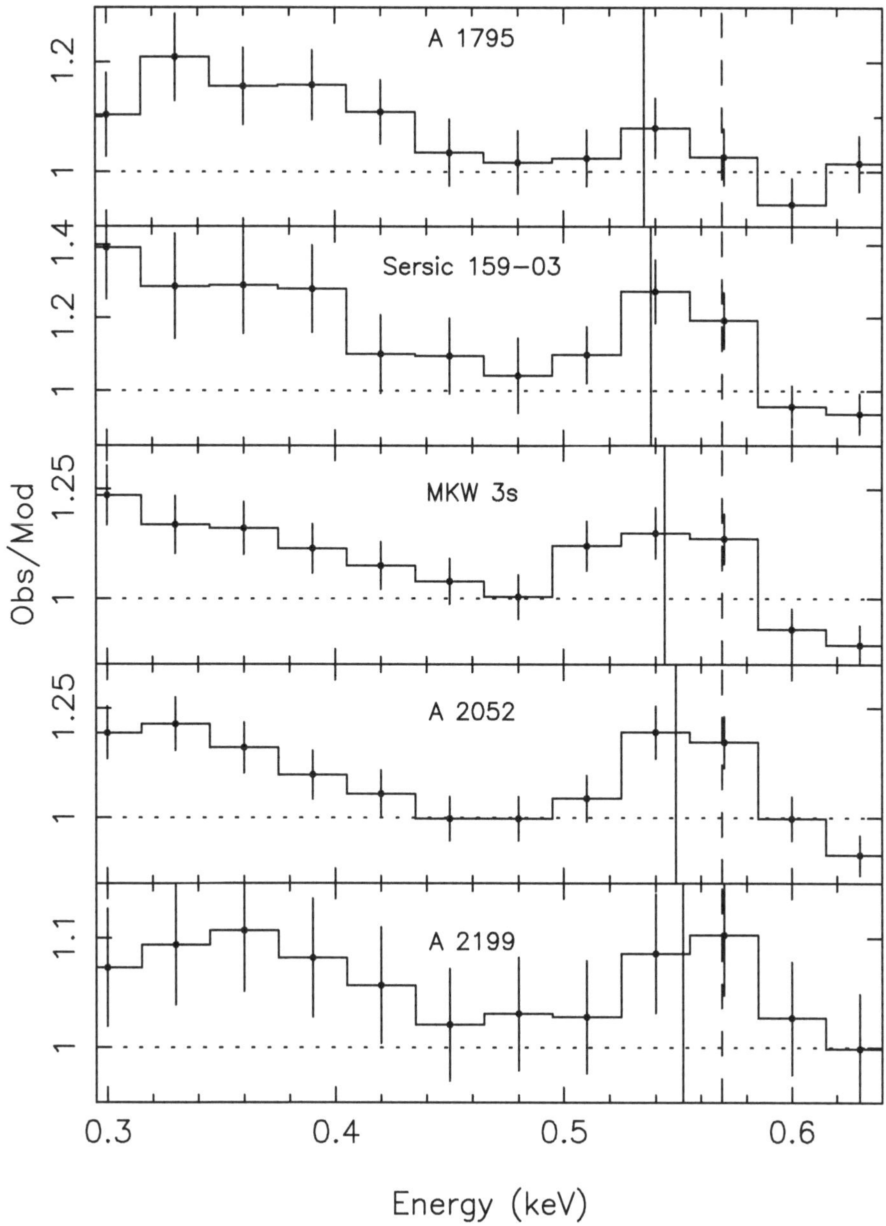

Figure 2. Fit residuals with respect to a two temperature model for the outer 4–12 arcmin part of six clusters. The position of the O VII triplet in the cluster restframe is indicated by a solid line and in our Galaxy's rest frame by a dashed line at 0.569 keV (21.80 Å). The fit residuals for all instruments (MOS, pn) are combined. The instrumental resolution at 0.5 keV is ~60 eV (FWHM).

temperature and density, and it is here that we expect to see most of the X-ray emission of the warm gas.

4. Properties of the warm gas

The temperature of the warm gas that we find for all our clusters is 0.2 keV. The surface brightness of the warm gas within the field of view of the XMM-Newton telescopes is approximately constant, with only a slight enhancement towards the center for some clusters (a stronger increase towards the center for Sérsic 159-03 is discussed in the next section). In Table 1 we list the central surface brightness S_0 as estimated from our XMM-Newton data, expressed as the emission measure per solid angle. We use $H_0 = 70$ km s^{-1} Mpc^{-1} throughout this paper. Using the known angular distance to the cluster we estimate the same quantity in units of m^{-5} (see also Table 1). We then use a simplified model for the geometry of the emitting warm gas, namely a homogeneous sphere with uniform density. The radius R of this sphere is estimated from the radial profile of the Rosat PSPC 1/4 keV profile around the cluster, and is also listed in Table 1. We find typical radii of 2–6 Mpc, i.e. the emission occurs on the spatial scale of a supercluster. From this radius and the emission measure, the central hydrogen density $n_{\rm H}(0)$ is estimated. We find typical densities of the order of 50-150 m^{-3}. Assuming a different density profile (for example $n(r) = n(0)[1 + (r/a)^2]^{-1}$ for $r < R$ and $n = 0$ for $r > R$) yields central densities that are only 20–50 % larger. These densities are 200–600 times the average baryon density of the universe. We also estimate the total hydrogen column density, which is typically $1.6 - 2.8 \times 10^{25}$ m^{-2}. Using then the measured oxygen abundances from our XMM-Newton data (essentially determined by the ratio of the O VII triplet to the soft X-ray excess), which are typically 0.1 times solar, we then derive total O VII column densities of the order of $0.4 - 0.9 \times 10^{21}$ m^{-2}. These column densities and the typical sizes of the emitting regions are similar to those as calculated for the brightest regions in the simulations of Fang et al. (2002). We have taken here a solar oxygen abundance of 8.5×10^{-4} and an O VII fraction of 32 %, corresponding to a plasma with a temperature of 0.2 keV.

Finally, we determined the total mass of the warm gas (M_w, Table 1). This mass is for most clusters comparable to the total cluster mass M_A within the Abell radius (2.1 Mpc for our choice of H_0) as derived by Reiprich & Böhringer (2002) for the same clusters.

We make here a remark on MKW 3s and A 2052. These clusters are separated by only 1.4 degree and both belong to the southernmost extension of the Hercules supercluster (Einasto et al. 2001). A 2199, at 35 degrees to the North, is at the northernmost end of the same supercluster. The redshift distribution of the individual galaxies in the region surrounding MKW 3s ($z = 0.046$) and

Table 1. Properties of the warm gas

Parameter	A 1795	Sérsic 159−03	MKW 3s	A 2052	A 2199
Redshift	0.064	0.057	0.046	0.036	0.030
Scale (kpc/arcmin)	71	64	52	41	35
S_0 [a]	49±37	71±41	84±18	73±15	28±13
S_0 [b]	10	18	33	46	24
R (arcmin)	80	60	60	80	60
R (Mpc)	5.7	3.8	3.1	3.3	2.1
$n_{\rm H}(0)$ ($\rm m^{-3}$)	45	80	120	140	120
$N_{\rm H}(0)$ (10^{24} $\rm m^{-2}$)	16	19	23	28	16
Abundance O	0.08±0.05	0.08±0.03	0.09±0.03	0.12±0.03	0.16±0.05
$N_{OVII}(0)$ (10^{20} $\rm m^{-2}$)	4	4	6	9	7
M_w (10^{15} $\rm M_\odot$)	1.2	0.6	0.5	0.7	0.2
$M_{\rm A}$ (10^{15} $\rm M_\odot$)	1.1	0.5	0.5	0.4	0.6

[a] Surface brightness, expressed as emission measure per solid angle (10^{68} $\rm m^{-3}arcmin^{-2}$).
[b] Surface brightness in 10^{26} $\rm m^{-5}$.

A 2052 ($z = 0.036$) shows two broad peaks centered around the redshifts of these clusters, but galaxies with both redshifts are found near both clusters. Therefore this region has a significant depth (43 Mpc) as compared to the projected angular separation (4.3 Mpc). The relative brightness of the warm gas near these clusters (as seen for example from the value of S_0) is then explained naturally if there is a filament connecting both clusters. In that case we would see the filament almost along its major axis.

5. Non-thermal emission?

Apart from the large scale, extended emission from the warm gas some clusters also exhibit a centrally condensed soft excess component (Fig. 3). In MKW 3s, A 2052 and A 2199 this central enhancement is relatively weak. It could be a natural effect of the enhanced filament density close to the cluster centers. In A 1795 and Sérsic 159−03 the enhancement is much larger. It is unlikely that this emission component for the latter two clusters also originates from projected filaments in the line of sight - the high surface brightness would necessitate filaments of length far larger than a cluster's dimension. Another possibility is warm gas within the cluster itself. In order to avoid the rapid cooling which results from this gas assuming a density sufficient to secure pressure equilibrium with the hot virialized intracluster medium, it should be be magnetically isolated from the hot gas. Yet another viable model for the central soft component is non-thermal emission. We note that the soft excess in the center of Coma and A 3112 also possibly has a non-thermal origin, as there is no clear evidence for oxygen line emission in their spectra. Clearly, deeper spectra and in particular a higher spectral resolution is needed to discriminate

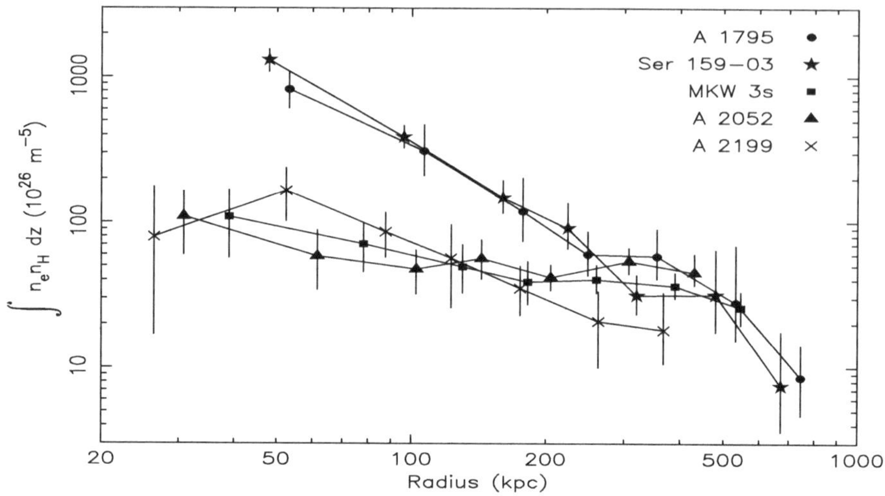

Figure 3. Emission measure integrated along the line of sight for five clusters of galaxies.

models. At this conference, several new mission concepts have been presented that may resolve these issues in the near future.

Acknowledgments This work is based on observations obtained with XMM-Newton, an ESA science mission with instruments and contributions directly funded by ESA Member States and the USA (NASA). SRON is supported financially by NWO, the Netherlands Organization for Scientific Research.

References

Cen, R., & Ostriker, J.P. 1999, ApJ, 514, 1

Einasto, M., Einasto, J., Tago, E., Müller, V., & Andernach, H. 2001, AJ, 122, 2222

Fang, T., Bryan, G.L., & Canizares, C.R., 2002, ApJ, 564, 604

Kaastra, J.S., Lieu, R., Tamura, T., Paerels, F. B. S., & den Herder, J. W., 2003a, A&A, 397, 445

Kaastra, J.S., Tamura, T., Peterson, J.R., et al., 2003b, A&A, submitted

Kaastra, J. S., Lieu, R., Tamura, T., Paerels, F. B. S., & den Herder, J. W., 2003c, Adv. Sp. Res., in press

Lieu, R., Mittaz, J.P.D., Bowyer, S., et al. 1996a, ApJ, 458, L5

Lieu, R., Mittaz, J.P.D., Bowyer, S., et al. 1996b, Science 274, 1335

Nevalainen, J., Lieu, R., Bonamente, M., & Lumb, D., 2003, ApJ, 584, 716

Peterson, J.R., Kahn, S.M., Paerels, F.B.S., et al., 2003, ApJ, in press

Reiprich, T.H., & Böhringer, H., 2002, ApJ, 567, 716

Sarazin, C.L., & Lieu, R., 1998, ApJ, 494, L177

Snowden, S.L., Freyberg, M.J., Plucinsky, P.P., et al., 1995, ApJ, 454, 643

XMM-NEWTON DISCOVERY OF AN X-RAY FILAMENT IN COMA

A. Finoguenov[1], U.G. Briel[1] and J.P. Henry[2]

[1]*Max-Planck-Institut fuer extraterrestrische Physik, Giessenbachstraße, 85748 Garching, Germany*

[2]*Institute for Astronomy, University of Hawaii, 2680 Woodlawn Drive, Honolulu, Hawaii 96822, USA*

Abstract XMM-Newton observations of the outskirts of the Coma cluster of galaxies confirm the existence of warm X-ray gas claimed previously and provide a robust estimate of its temperature ($\sim$ 0.2 keV) and oxygen abundance ($\sim$ 0.1 solar). Associating this emission with a 20 Mpc infall region in front of Coma, seen in the skewness of its galaxy velocity distribution, yields an estimate of the density of the warm gas of $\sim 50\rho_{critical}$. Our measurements of gas mass associated with the warm emission strongly support its nonvirialized nature, suggesting that we are observing the warm-hot intergalactic medium (WHIM). Our measurements provide a direct estimate of the O, Ne and Fe abundance of the WHIM. Differences with the reported Ne/O ratio for some OVI absorbers hints to different origin of the OVI absorbers and the Coma filament.

Keywords: clusters: individual: Coma — cosmology: observations — intergalactic medium — large-scale structure

Introduction

The total amount of baryons found in the local Universe accounts only for 10% of the values implied from the primordial nucleosynthesis, observed at high-redshift (Fukigita, Hogan, Peebles 1998 and references therein) and within the local closed box systems, such as clusters of galaxies (White, Briel, Henry 1993). It has been predicted by Cen & Ostriker (1999) that most of the missing baryons reside in the warm-hot intergalactic medium (WHIM), owing to the shocks driven by the large-scale structure formation that results in an order of magnitude drop in the star-formation rate density (Nagamine et al. 2001).

Since then, several surveys have been undertaken to find the WHIM. Tripp, Savage, Jenkins (2000) have claimed a substantial fraction of the missing

R. Lieu and J. Mittaz (eds.), Soft X-Ray Emission from Clusters of Galaxies and Related Phenomena, 45–52.

baryons in the form of OVI absorbers. However, their observations lack information on the ionization equilibrium as well as O abundance required to link the abundance of OVI ions to underlying O mass and then to the amount of baryons. First Chandra grating results (Nicastro et al. 2002, Mathur, Weinberg, Chen 2002) on detection of OVII absorption lines allowed, for the first time, an estimate of the ionization state of the gas, suggesting that OVI absorbers account for over 80% of the local baryons. Still, such claims rely on the assumed O abundance.

Searches for the warm X-ray emission with ROSAT, have found some association of soft X-ray emission with galactic filaments (Scharf et al. 2000; Zappacosta et al. 2002). Based on the prediction that hotter gas will be in denser regions, search for the X-ray emission from the regions connecting clusters of galaxies has been a separate, although related, topic. Briel & Henry (1995) set upper limits with stacked analysis of ROSAT all-sky survey data on the emitting gas density to be less than $7.4 \times 10^{-5} h^{1/2}$ cm^{-3}, assuming a temperature of 0.5 keV and an iron abundance of 0.3 solar.

In this *Contribution* we carry out a spectroscopic study of the warm X-ray gas in the outskirts of the Coma cluster, using the mosaic of observations carried out by the XMM-Newton. We adopt the Coma cluster redshift of 0.023, $H_o = 70$ km s^{-1} Mpc^{-1}, and quote errorbars at the 68% confidence level. One degree corresponds to 1.67 Mpc.

1. Observations

The initial results of the XMM-Newton performance verification observations of the Coma cluster are reported in Briel et al. (2001 and references therein). In addition to the observations reported there, three additional observations have been carried out, completing the planned survey of the Coma cluster.

The Coma observations reported here are performed with EPIC pn detector, using its medium filter, which leads to a different detector countrate from the cosmic and galaxy emission in the soft band, compared to the usual thin filter. We used a 100 ksec observation of the point source APM08279+5255, performed with the same (medium) filter, to measure these two diffuse emission components. To ensure the compatibility of the X-ray background, we excise from the spectral extraction region the APM08279+5255 quasar and the X-ray sources in Coma identified with the Coma galaxies (which otherwise increase the apparent CXB flux by 10%). Since there is a mismatch in the N_H toward the Coma and the APM fields, we use an additional background observation, obtained with the filter wheel closed, to extract the foreground (the Local Bubble and the Milky Way halo, e.g. Lumb et al. 2002) and CXB spectra in APM field. We then use the same spectral shape and normalization, but change the

N_H from $3.9{\pm}0.3\times10^{20}$ cm^{-2} (APM) to 9×10^{19} cm^{-2} in the direction of the Coma cluster. Since some foreground components, like the Local Bubble, are not subject to the absorption, we introduce a separate unabsorbed component of 0.07 keV temperature (Lumb et al. 2002) in both the Coma and the APM fields. We have also checked that the derived results on the soft X-ray emission from Coma do not depend on the possible large-scale variation in the intensity of the 0.07 keV component.

To improve on the detection statistics of the warm emission and to reduce the systematic effects of subtraction of the cluster emission, we concentrate on the *outskirts* of the Coma cluster. In Fig.1 we show the location of the spectral extracting areas. We select regions $\sim 40'$ from the Coma center to the North-West, North, North-East, South and South-East. The South-West direction is complicated by the presence of the infalling subcluster (NGC4839), while some other pointings were affected by background flares, which after screening lead to insufficient exposures.

2. Results

Our major results are shown in Fig.2 and listed in Table 1. We find that the soft excess in Coma on spatial scales of $30' - 50'$ is characterized by thermal plasma emission, as indicated by our discovery of O lines, with a characteristic temperature of ~ 0.2 keV, varying with position on the sky. The highest temperature of the warm component corresponds to the hot spot in Coma, found in ASCA observations (Donnelly et al. 1999), and identified as an accreting zone of the Coma cluster.

At the moment, it is difficult from the O line redshift alone to decide whether this emission is galactic or extragalactic. However, some apparent differences with the galactic emission are already noticeable. The soft component is centered on Coma and has an amplitude exceeding the variation of the underlying galactic emission (Bonamente, Joy, Lieu 2003). An O abundance of 0.1 times solar is much lower than the galactic value of one solar (Markevitch et

Table 1. Characteristics of the warm emission around Coma

Location (Field)	kT_{Coma} keV	kT_{warm} keV	$Z/Z_\odot^\dagger$	$\rho^\flat/\rho_{crit}$	$M_{gas}^\flat$ $10^{12}M_\odot$
Coma–0	15 ± 6	0.19 ± 0.01	0.06 ± 0.02	65 ± 8	4.6 ± 0.5
Coma–3	5.8 ± 0.9	0.19 ± 0.02	0.04 ± 0.02	73 ± 10	4.5 ± 0.6
Coma–7	10 ± 3	0.22 ± 0.02	0.07 ± 0.03	40 ± 11	2.4 ± 0.6
Coma–11	16 ± 2	0.24 ± 0.01	0.09 ± 0.01	85 ± 9	6.0 ± 0.1
Coma–13	3.1 ± 0.5	$0.17{\pm}0.01$	0.12 ± 0.04	54 ± 12	2.4 ± 0.6

$\flat$ Estimated by assuming a 90 Mpc distance, projected length of 20 Mpc, area of the extraction regions in Fig.1 and $n_e = 5.5\times10^{-7}$ cm^{-3} in correspondence to ρ_{crit} for h=0.7

† Assuming a solar abundance ratio.

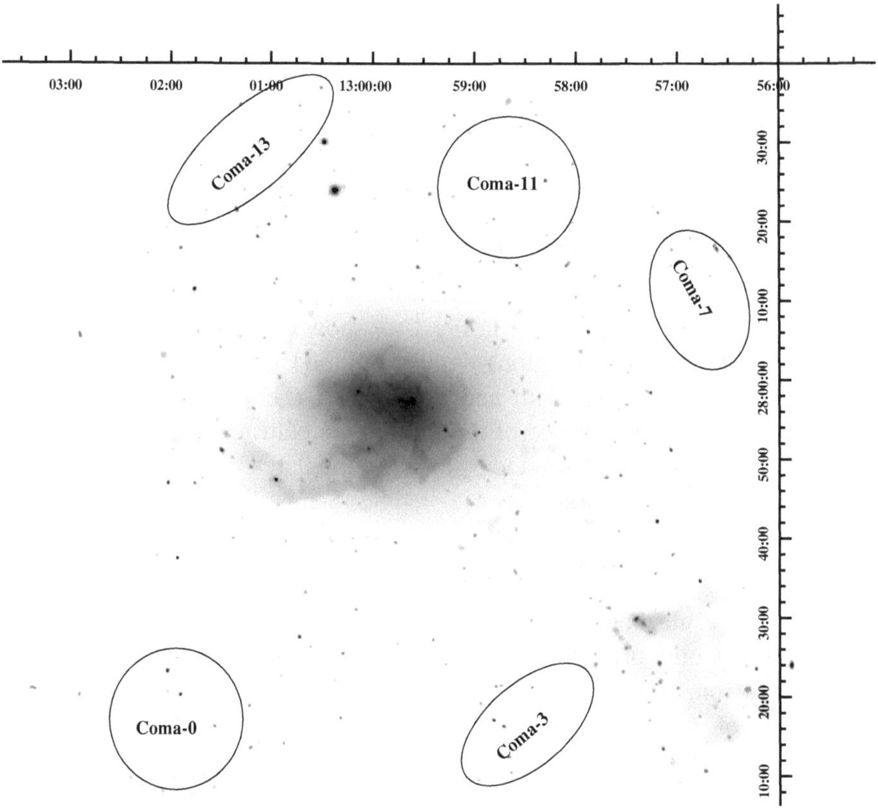

Figure 1. An image of the Coma cluster in the 0.5–2 keV band, showing the positions of the spectral extraction regions (solid ellipses) with names according to the XMM-Newton survey notation. The coordinate grid marks R.A., Dec. (J2000.0).

al. 2002; Freyberg & Breitschwerdt 2002). The temperature variations of the Coma warm component are not seen in the quiet zones of galactic emission at high latitudes. The observed warm component exceeds the level of the galactic emission by a factor of up to 5, so to significantly change the derived O abundance of the Coma warm emission a possible variation in the metallicity for the galactic emission should be by a factor of 3, which is not observed.

3. Interpretation

In the rest of this *Contribution* we present an interpretation assuming an extragalactic origin of the warm component. With the CCD-type spectrometry we can place limits on the possible redshifts of the warm component, which for the Coma-11 field is determined as $0.007 \pm 0.004 \pm 0.015$ (best-fit, statistical and systematic error). Any extragalactic interpretation should therefore con-

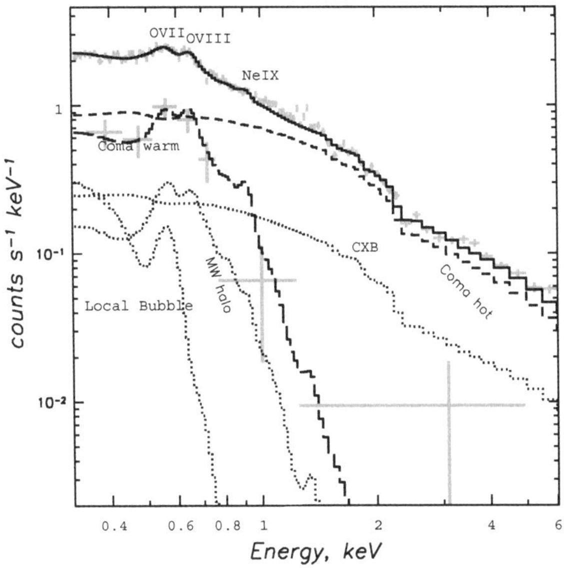

Figure 2. Detailed decomposition of the pn spectrum (small grey crosses) in the Coma-11 field into foreground and background components, obtained from the analysis of the APM field, plus hot and warm emission from the Coma cluster. Large crosses is a result of the in-field subtraction of Coma emission and detector background, possible due to differences in the spatial distributions between the Coma warm, Coma hot and detector background components.

centrate on the large-scale structure in front of the Coma cluster. We have carried out an analysis of the CfA2 galaxy catalog (Huchra, Geller, Corwin 1995) towards the Coma cluster, excising the South-West quadrant, where strong influence of the infalling subcluster NGC4839 on the velocity distribution has been suggested (Colles & Dunn 1996). Fig.3 illustrates our result, indicating a significant galaxy concentration in front of the Coma cluster, with velocities lying in the $4500 - 6000$ km s^{-1} range. The sky density of the infall zone is shown in the insert of Fig.3. It was constructed from the number of galaxies in the $4500 - 6000$ km s^{-1} velocity range minus the number of galaxies in the $8500 - 10000$ km s^{-1} bin, corresponding to a similar difference in the velocity dispersion from the cluster mean. By excising the galaxies at velocities lower than 6000 km s^{-1}, we recover a velocity dispersion of $\sim$ 700 km s^{-1}. Using the $M - \sigma$ relation of Finoguenov, Reiprich, Boehringer (2001), we find that such velocity dispersion is more in accordance with mass of the Coma cluster. The excess of galaxies exhibits a flat behavior within 0.8 degrees (1.3 Mpc), followed by a decline by a factor of 5 within 1.7 degrees (2.8 Mpc) and a subsequent drop by two orders of magnitude within 10 degrees (16.7 Mpc) The soft emission from the Coma is detected to a 2.6 Mpc distance from the

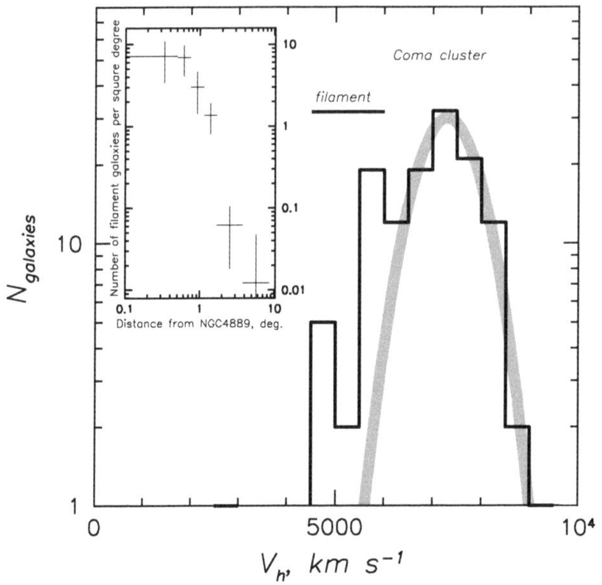

Figure 3. Galaxy distribution in the direction to the Coma cluster from CfA2 survey. The central 2 degrees in radius field centered on NGC4889 is shown, excluding the 60 degrees cone in the South-West direction from the center, to avoid the effect of the NGC4839 subcluster. An excess of galaxies over the Gaussian approximation of the velocity dispersion of the Coma cluster is seen in the $4500 - 6000$ km s^{-1} velocity range. The insert shows the spatial distribution of the excess galaxies in the $4500 - 6000$ km s^{-1} velocity bin over the galaxies in the $8500 - 10000$ km s^{-1} velocity bin, this time out to 10 degrees.

center (Bonamente et al. 2003) in a remarkable correspondence to the galaxy filament.

In the absence of the finger of god effect, validated below by our conclusion on the filamentary origin of this galaxy concentration, the infall zone is characterized by a 20 Mpc projected length. This length assumption affects the density estimates, while the O abundance is only based on the assumption of collisional equilibrium. We have verified the later assumption by studying the ionization curves for OVII and OVIII presented in Mathur et al. (2002). We have concluded that the assumption of pure collisional ionization is valid for our data, since $n_e > 10^{-5}$ cm^{-3} and $T > 2 \times 10^6$ K (0.17 keV). An advantage of our measurement is that we also determine the temperature by the continuum. Lower temperatures, which at densities near 10^{-5} cm^{-3} result in similar line ratios for OVII and OVIII, fail to produce the observed (H-e and He-e) bremsstrahlung flux at 0.7–1 keV.

An important question we want to answer is whether the detected emission originates from a group or from a filament. We cannot decide from the overdensity of the structure ($\sim 50\rho_{crit}$), as it suits both. However, the implied mass of the structure exceeds the mass of the Coma cluster (see also Bonamente et al. 2003). On the other hand the temperature of the emission is ~ 0.2 keV, almost two orders of magnitude lower than that of the Coma cluster. It takes a few hundred groups with virial temperature of 0.2 keV to make up a mass of the structure, which leads to an overlapping virial radii if we are to fit them into the given volume of the structure. Thus we conclude that the observed structure is a filament.

For the Coma-11 field, where the statistics are the highest, we investigated the effect of relaxing the assumption of solar abundance ratios. We note that an assumption for C abundance is important for overall fitting and a solar C value would affect our results. When left free, however, the C abundance tends to go to zero. Also, there is no systematic dependence of our results on the assumed C abundance as long as the C/O ratio is solar or less, which is the correct assumption from the point of view of chemical enrichment schemes and observations of metal-poor stars in our Galaxy. Significant abundance measurements are: O/O$_\odot$ $= 0.14 \pm 0.02$, Ne/Ne$_\odot$ $= 0.14 \pm 0.06$, Fe/Fe$_\odot$ $= 0.04^{+0.03}_{-0.01}$ (assuming Fe$_\odot$$/H = 4.26 \times 10^{-5}$ by number), indicating that Fe is underabundant by a factor of three in respect to the solar Fe/O ratio, implying a dominant contribution of SN II to Fe enrichment. Element abundances for the hot emission obtained from our XMM data on the center of Coma are Mg/Mg$_\odot$ $= 0.36 \pm 0.12$, Si/Si$_\odot$ $= 0.45 \pm 0.08$, S/S$_\odot$ $= 0.01 \pm 0.01$, Fe/Fe$_\odot$ $= 0.20 \pm 0.03$, also suggesting prevalence of SN II in Fe enrichment, although Fe abundance is a factor of 5 higher compared to the filament. The filament Ne/O ratio reveals a subtle difference with the OVI absorbers. Nicastro et al. (2002) reports Ne/O $=$ 2 times solar, while we observe a lower (solar) ratio at 95% confidence. Lower ratios seem to be a characteristic of enrichment at $z > 2$ (Finoguenov, Burkert, Boehringer 2003), while WHIM originates at $z < 1$ (Cen & Ostriker 1999). While we would rather have more observational evidence on the dispersion of O to alpha element ratios, we believe that this is a major source of systematics in interpreting the element abundance of X-ray filaments as a universal value.

To illustrate the point, we calculate the amount of baryons traced by the OVI absorbers:

$$\Omega_b^{OVI} = 0.020 h_{70}^{-1} \left(\frac{0.14}{10^{[O/H]}} \right) \left(\frac{\langle [f(OVI)]^{-1} \rangle}{32} \right) \quad (1)$$

where the original value of $\Omega_b(OVI) = 0.0043 h_{70}^{-1}$ of Tripp et al. (2000) has been scaled for a ionization equilibrium implied by measurements of Mathur et al (2002) and using the measurements of the O abundance reported here. The formal errorbar is 0.006, mostly from the uncertainty in the estimate for ionization. If, on the other hand, our measurements of Ne abundance are used, and the Ne/O ratio for OVI absorbers is taken from Nicastro et al. (2002),

$$\Omega_b^{OVI} = 0.040 h_{70}^{-1} \left(\frac{0.14}{10^{[Ne/H]}} \frac{10^{[Ne/O]}}{2} \right) \left(\frac{\langle [f(OVI)]^{-1} \rangle}{32} \right) \qquad (2)$$

So, with $\Omega_b^{\rm total} = 0.039$, there is no room left for other major components of local baryons, Ly_α absorbers (0.012 ± 0.002) and stars and clusters of galaxies (~ 0.006; e.g., Finoguenov et al. 2003 and references therein). If O is depleted onto dust grains in the OVI absorbers, as suggested by Nicastro et al. (2002) as an explanation of the high Ne/O ratio, the solar Ne/O ratio in our observations simply indicates dust sputtering, then the baryon budget of OVI absorbers follows the calculation in Eq.2 and yields, consequently, unacceptably high values.

References

Bonamente, M., Joy, M., Lieu, R. 2003, ApJ, in press (astro-ph/0211439)

Briel, U.G., Henry, J.P. 1995, A&A, 302, 9

Briel, U.G., Henry, J.P., Lumb, D.H., Arnaud, M., Neumann, D., et al. 2001, A&A, 365, L60

Cen, R., Ostriker, J.P. 1999, ApJ, 514, 1

Colles, M., Dunn, A.M. 1996, ApJ, 458, 435

Donnelly, R.H., Markevitch, M., Forman, W., Jones, C., Churazov, E., Gilfanov, M. 1999, ApJ, 513, 690

Finoguenov, A., Reiprich T., Boehringer, H. 2001, A&A, 368, 749

Finoguenov, A., Burkert, A., Boehringer, H. 2003b, ApJ, submitted

Freyberg, M. J., Breitschwerdt, D. 2002, Proc. of JENAM 2002, in press

Fukugita, M., Hogan, C.J., Peebles, P.J.E. 1998, ApJ, 503, 518

Huchra, J.P., Geller, M.J., Corwin, H.G.Jr. 1995, ApJS, 99,391

Lumb, D.H., Warwick, R.S., Page, M., De Luca, A. 2002, A&A, 389, 93

Markevitch, M., Bautz, M.W., Biller, B., et al. 2002, ApJ, in press, (astro-ph/0209441)

Mathur, S., Weinberg, D.H., Chen, X. 2002, ApJ, in press, astro-ph/0206121

Nagamine, K., Fukugita, M., Cen, R., Ostriker, J.P. 2001, ApJ, 558, 497

Nicastro, F., Zezas, A., Drake, J., Elvis, M., Fiore, F., Fruscione, A., Marengo, M., Mathur, S., Bianchi, S. 2002, ApJ, 573, 157

Scharf, C., Donahue, M., Voit, G.M., Rosati, P., Postman, M. 2000, ApJ, 528, L73

Tripp, T.M., Savage, B.D., Jenkins, E.B. 2000, ApJ, 534, L1

White, S.D.M., Briel, U.G., Henry, J.P. 1993, MNRAS, 261, L8

Zappacosta, L., Mannucci, F., Maiolino, R., Gilli, R., Ferrara, A., Finoguenov, A., Nagar, N. M., Axon, D. J. 2002, A&A, 394, 7

THE CLUSTER ABELL 85 AND ITS X-RAY FILAMENT REVISITED BY CHANDRA AND XMM-NEWTON

Florence Durret[1], Gastão B. Lima Neto[2], William R. Forman[3], Eugene Churazov[4]

[1] *Institut d'Astrophysique de Paris, 98bis Bd Arago, 75014 Paris, France*

[2] *Instituto Astronômico e Geofísico/USP, Av. Miguel Stefano 4200, São Paulo/SP, Brazil*

[3] *Harvard Smithsonian Center for Astrophysics, 60 Garden St, Cambridge MA 02138, USA*

[4] *MPI für Astrophysik, Karl Schwarzschild Strasse 1, 85740 Garching, Germany*

Abstract We have observed with XMM-Newton the extended 4 Mpc filament detected by the ROSAT PSPC south east of the cluster of galaxies Abell 85. We confirm the presence of an extended feature and find that the X-ray emission from the filament is best described by thermal emission with a temperature of $\sim$ 2 keV. This is significantly lower than the ambient cluster medium, but is significantly higher than anticipated for a gas in a weakly bound extended filament. It is not clear whether this is a filament of diffuse emission or a chain of several groups of galaxies.

1. X-ray and galaxy filaments

Numerical simulations of structure formation in cold dark matter (CDM) cosmological models (e.g. Frenk et al. 1983, Jenkins et al. 1998) predict that galaxies form preferentially along filaments, and that clusters are located at the intersections of these filaments. The detection of filaments visible in X-rays can therefore be a test to scenarios of large scale structure formation.

Various types of observations indeed seem to indicate that clusters remember how they formed. Some clusters show preferential orientations at various scales; in Abell 85, for example, the central cD galaxy, the brightest galaxies, the overall galaxy distribution and the X-ray gas distribution all show elongations along the same direction (Durret et al. 1998b). At even larger scales, the probability for no alignment between two clusters separated by a few Mpc has been found to be very small (Chambers et al. 2002).

Bridges of material (galaxies and/or groups of galaxies) have been detected between several clusters (West et al. 1995, West & Blakeslee 2000), though

R. Lieu and J. Mittaz (eds.), Soft X-Ray Emission from Clusters of Galaxies and Related Phenomena, 53–61.

they are not very easy to spot since they require the measurements of numerous galaxy redshifts.

When ROSAT all sky survey background fluctuations were correlated with the Abell cluster catalogue, no filament was detected, implying an upper limit to the electron density consistent with that predicted by hydrodynamical simulations (Briel & Henry 1995). A few X-ray filaments or at least highly elongated X-ray emitting regions were detected however in a few clusters such as Coma (Vikhlinin et al. 1997), Abell 2125 (Wang et al. 1997), Abell 85 (Durret et al. 1998b), the core of the Shapley supercluster (Kull & Böhringer 1999), Abell 1795 (Fabian et al. 2001, Markevitch et al. 2001), and between Abell 3391 and Abell 3395 (Tittley & Henriksen 2001).

2. The X-ray filament in Abell 85

The complex of clusters Abell 85/87/89 is a well studied structure, dominated by Abell 85 at a redshift of $z = 0.056$ (Durret et al. 1998b) (at $z = 0.056$, 1 arcmin = $90h_{50}^{-1}$ kpc). Abell 87 is not detected in X-rays, neither by the ROSAT PSPC nor in our XMM-Newton image, so it is probably not a real cluster, but perhaps a concentration of several groups. This would agree with Katgert et al. (1996) who find two different redshifts in that direction, one coinciding with that of Abell 85 and the other at z=0.077. Abell 89 comprises two background structures and will not be discussed further (see Durret et al. 1998b for details).

The ROSAT PSPC image of Abell 85 has revealed an elongated X-ray structure to the southeast of the main cluster Abell 85. This filament is at least 4 Mpc long (projected extent on the sky), and it is not yet clear whether this is a filament of diffuse emission or a chain made by several groups of galaxies. Durret et al. (1998b) have shown that after subtracting the modelled contribution of Abell 85 the remaining emission has the appearance of being made of several groups, each having an X-ray luminosity of about 10^{42} erg s^{-1}. Kempner et al. (2002) have recently observed Abell 85 with Chandra; they present data concerning the X-ray concentration at the northernmost end of the filamentary structure, but the filament itself is outside their field.

We have reobserved Abell 85 with XMM-Newton. The cleaned, background subtracted and exposure map corrected merged MOS1+2 image is shown in Fig. 1. The image obtained after subtracting an azimuthal average of the X-ray emission of the overall cluster is shown in Fig. 2. Besides the main cluster and the south blob the filament is clearly visible towards the south east.

Events were extracted inside an elongated region of elliptical shape centered at R.A. = $0^{h}42^{m}15.0^{s}$ and Decl. = $-09°39'24''$ (J2000), with major and minor axis of 8 and 4 arcmin, respectively. Three circular regions corresponding to point sources (see Fig. 2) were excluded.

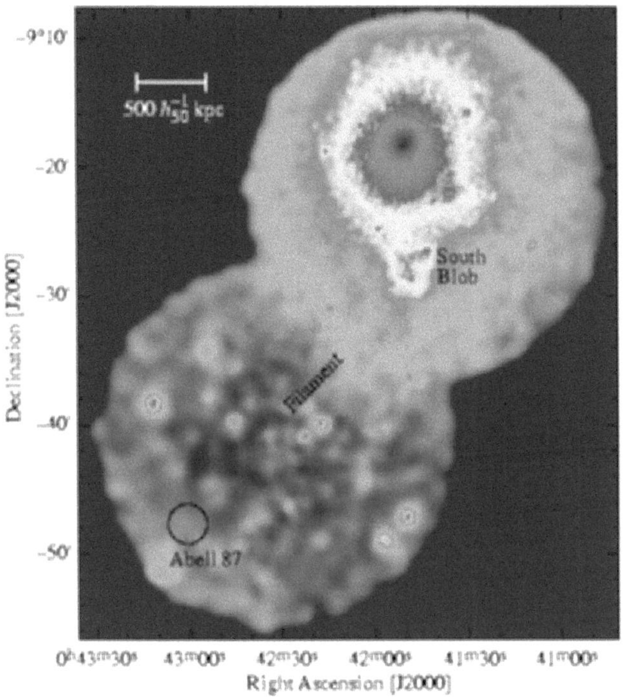

Figure 1. Cleaned, background subtracted, exposure map corrected image obtained from the two MOS1 and MOS2 XMM-Newton exposures. The image has been smoothed with a Gaussian of $\sigma = 5$ arcsec.

The background was taken into account by extracting spectra (for MOS1 and MOS2) from the EPIC blank sky templates described by Lumb et al. (2002). We have applied the same filtering procedure to the background event files and extracted the spectra in the same elliptical region of the filament in detector coordinates. Finally, the spectra have been rebinned so that there are at least 30 counts per energy bin.

The count rates after background subtraction are 0.1096 ± 0.0055 cts/s and 0.0899 ± 0.0057 cts/s for the MOS1 and MOS2 respectively, corresponding to total source counts of 1354 and 1115. The spectral fits were done with XSPEC 11.2, with data in the range [0.3–10 keV], simultaneously with MOS1 and MOS2. A MEKAL model (Kaastra & Mewe 1993, Liedahl et al. 1995) with photoelectric absorption given by Balucinska-Church & McCammon (1992) was used to fit the spectral data. Since the spectra were rebinned, we have used standard χ^2 minimization. Fig. 3 shows the MOS1 and MOS2 spectra, together with the best MEKAL fits and residuals.

Table 1 summarizes results of the spectral fits for the filament. If $N_{\rm H}$ is left free to vary in the fit, it is not well constrained, since its 90% confidence

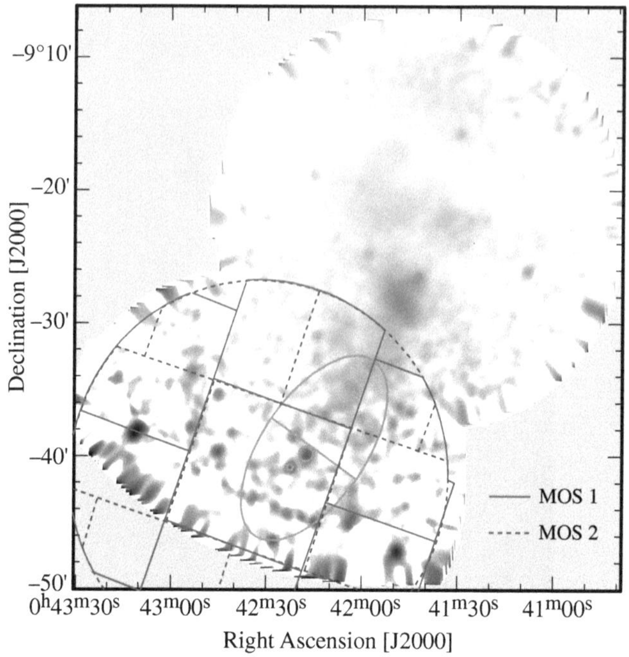

Figure 2. Merged XMM-Newton image obtained after subtracting an azimuthal average of the X-ray emission of the overall cluster. The ellipse, excluding the three circles, shows the region where we extracted the events for the spectral analysis. The straight line dividing the ellipse in two defines the "north" and "south" halves of the filament. The borders of MOS1 and MOS2 are also shown (thin lines).

Table 1. Results of the spectral fits for the filament with a MEKAL model. Error bars are 90% confidence limits.

kT [keV]	Z [solar]	N_H [10^{20} cm^{-2}]	χ^2/dof
$1.9^{+0.5}_{-0.4}$	$0.05^{+0.13}_{-0.05}$	$8.2^{+4.1}_{-3.6}$	194.0/192
$2.4^{+0.5}_{-0.3}$	$0.17^{+0.17}_{-0.12}$	3.16 (fixed)	199.5/193

interval is $(4.6 \leq N_{\rm H} \leq 12.3) \times 10^{20}\,{\rm cm}^{-2}$, and the corresponding X-ray temperature is in the range $1.5 \leq kT \leq 2.4\,{\rm keV}$. For the metallicity, we only have an upper limit $Z \leq 0.18\ Z_{\odot}$, which is almost the same as the mean metallicity obtained in the fit with fixed $N_{\rm H}$.

The galactic neutral hydrogen column density in the direction of the filament is $N_{\rm H} = 3.16 \times 10^{20}\,{\rm cm}^{-2}$ (Dickey & Lockman 1990). When $N_{\rm H}$ is fixed to this value, a MEKAL fit to the filament spectrum gives a gas temperature of $2.0 \leq kT \leq 2.8\,{\rm keV}$ and a metallicity of $0.04 \leq Z \leq 0.33 Z_{\odot}$ (90% confidence range).

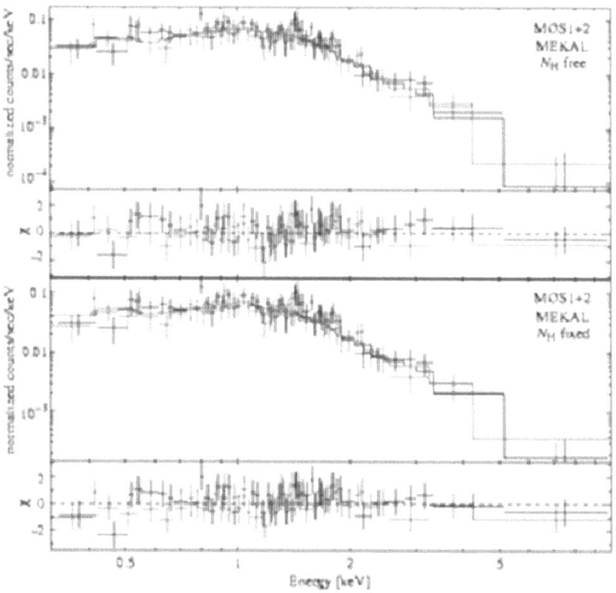

Figure 3. XMM-Newton MOS1 and MOS2 rebinned spectra of the filament region with the best MEKAL fits superimposed. Top: fit with the hydrogen column density free. Bottom: fit with $N_{\rm H}$ fixed at the galactic value at the position of the filament.

We also attempted to fit a power-law to the spectra. The resulting photon index is in the interval 2.5–3.0 and 1.8–2.0 when $N_{\rm H}$ is left free to vary or is fixed, respectively. The MEKAL and power law fits are of comparable statistical quality when $N_{\rm H}$ is left free to vary; however, when $N_{\rm H}$ is fixed, the MEKAL fit is significantly better. Since the fitted value of $N_{\rm H}$ for the power law is unreasonably large and the value of χ^2 unacceptable when $N_{\rm H}$ is fixed at the Galactic value, we reject the power law as a possible spectral model and favour thermal emission.

Visual inspection of the filament (Fig. 2) reveals a North → South gradient. We defined two halves of the filament region and analysed them in the same way as the whole ellipse. Within error bars, we find no gradient of temperature or metal abundance.

3. A more general X-ray view of Abell 85

The XMM temperature map of Abell 85 is displayed in Fig. 4. The three expected cool regions (the central cD where a cooling flow is observed, the south clump and the southwest clump) are clearly seen. The region just above the southern clump is hot, in agreement with Kempner, Sarazin & Ricker (2002), and with the ASCA temperature map of Donnelly et al. (2003, in preparation).

This agrees with the idea that in the impact region where material from the filament is falling onto Abell 85 a shock may occur and the gas is compressed and hotter (Durret et al. 1998b).

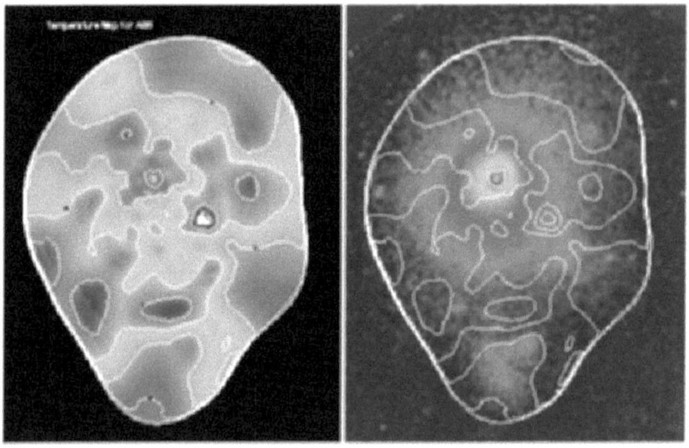

Figure 4. Left: temperature map with iso-temperature contours superimposed. Lower temperatures are 4–5 keV (lighter) and highest temperatures are $\sim$ 7 keV (darker). The map is adaptively smoothed. Right: central part of Fig. 1 with the iso-temperature contours from the temperature map superimposed.

Fig. 5 shows the radial profiles of temperature and metal abundance for the overall cluster, computed in circular rings around the cluster center both from Chandra and Beppo-Sax data. Results appear quite different when the hydrogen column density is fixed (to the Lockman & Dickey 1990 value) or when it is left free to vary. A clear temperature drop and metallicity rise are observed towards the center.

4. Discussion and conclusions

Our XMM-Newton observations confirm the existence of a highly elongated filamentary like structure extending from the South blob to the south east of Abell 85 along the direction defined by all the structures pointed out by Durret et al. (1998b). The fact that the spatial structure of the X-ray filament detected by XMM-Newton cannot be exactly superimposed to that obtained from ROSAT data shows that it is still difficult to determine exactly its structure.

However, we have shown that the X-ray spectrum from this structure is most likely thermal and its temperature is about 2.0 keV, consistent with that of groups. This value is notably cooler than that of the main cluster: the temperature map by Markevitch et al. (1998) shows the presence of gas at about 5 keV in the region at a distance from the cluster center at least as far as the northern part of the ellipse. So, we appear to be seeing cool gas as it enters the cluster core. A full description of the X-ray properties of the filament can be found in Durret et al. (2003).

In an attempt to characterize substructure in the filament region from the galaxy distribution, we have analysed the galaxy distribution in this zone. There are 18 galaxies in our redshift catalogue with redshifts in the cluster velocity range (Durret et al. 1998a) and coinciding with the position of the X-ray filament. The velocity histogram of these 18 galaxies shows a pair of galaxies with velocities around 14060 km s^{-1} and two peaks, both including 8 galaxies: the first one is centered on 15770 km s^{-1} with a dispersion of 235 km s^{-1}, and the second one is centered on 17270 km s^{-1} with a dispersion of 380 km s^{-1}. However, neither of these two "groups" of galaxies is concentrated on the sky. Instead, their galaxy positions cover the entire filament, so we cannot say we are detecting any structure in the galaxy velocity distribution.

Acknowledgments

We acknowledge help with the XMM-Newton SAS software from Sébastien Majerowicz and Sergio Dos Santos. F.D. and G.B.L.N. acknowledge financial support from the USP/COFECUB. G.B.L.N. acknowledges support from FAPESP and CNPq. W. Forman thanks the Max-Planck-Institute für Astrophysik for its hospitality during the summer of 2002 and acknowledges support from NASA Grant NAG5-10044. This work is based on observations obtained with XMM-Newton, an ESA science mission with instruments and contributions directly funded by ESA Member States and the USA (NASA).

References

Balucinska-Church M. & McCammon D. 1992, ApJ 400, 699

Briel U. & Henry P. 1995, A&A 302, L9

Chambers S.W., Melott A.L., Miller C.J. 2002, ApJ 565, 849

Dickey J.M. & Lockman F.J. 1990, Ann. Rev. Ast. Astr. 28, 215

Durret F., Felenbok P., Lobo C. & Slezak E. 1998a, A&A Suppl. 129, 281

Durret F., Forman W., Gerbal D., Jones C. & Vikhlinin A. 1998b, A&A 335, 41

Durret F., Lima Neto G.B., Forman W., Churazov E. 2003, A&A Letters in press, astro-ph/0303486

Fabian A.C., Sanders J.S., Ettori E. et al. 2001, MNRAS 321, L33

Frenk C.S., White S.D.M. & Davis M. 1983, ApJ 271, 417

Jenkins A., Frenk C.S., Pearce F.R. et al. 1998, ApJ 499, 20

Kaastra J.S. & Mewe R. 1993, A&AS 97, 443

Katgert P., Mazure A., Perea J. 1996, A&A 310, 8

Kempner J.C., Sarazin C.L. & Ricker P.M. 2002, ApJ 579, 236

Kull A., Böhringer H. 1999, A&A 341, 23

Liedahl D.A., Osterheld A.L. & Goldstein W.H. 1995, ApJ 438, L115

Lumb D.H., Warwick R.S., Page M. & De Luca A. 2002, A&A 389, 93

Markevitch M., Forman W.R., Sarazin C.L., Vikhlinin A. 1998 ApJ, 503, 77

Markevitch M., Vikhlinin A., Mazzotta P. 2001 ApJ, 562, L153

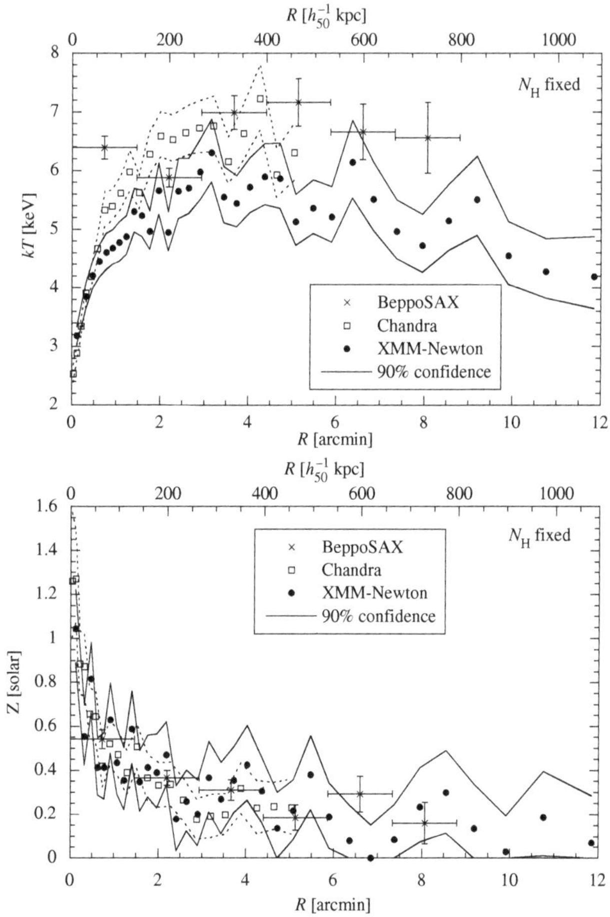

Figure 5. XMM-Newton (black circles and full lines), Chandra (empty squares and dotted lines) and Beppo-Sax (x's with error bars) temperature (top) and metallicity (bottom) profiles. All the spectral fits were done with the hydrogen column density fixed.

Tittley E.R. & Henriksen M. 2001, ApJ 563, 673
Vikhlinin A., Forman W. & Jones C. 1997, ApJL 474, L7
Wang Q.D., Connolly A. & Brunner R. 1997, ApJ 487, L13
West M.J., Jones C. & Forman W. 1995, ApJL 451, L5
West M.J. & Blakeslee J.P. 2000, ApJ 543, L27

II

ABSORPTION LINE STUDIES OF OUT GALAXY AND THE WHIM

CHANDRA DETECTION OF X-RAY ABSORPTION FROM LOCAL WARM/HOT GAS

T. Fang[1,2], C. Canizares[2], K. Sembach[3], H. Marshall[2], J. Lee[2], D. Davis[4]

[1]*Dept. of Physics,Carnegie Mellon Univ., 5000 Forbes Ave., Pittsburgh, PA 15213*

[2]*MIT, Center for Space Research, 70 Vassar St., Cambridge, MA 02139*

[3]*STScI, 3700 San Martin Drive, Baltimore, MD 21218*

[4]*GSFC, Code 661., GLAST SSC, Greenbelt, MD 20771*

Abstract Recently, with the *Chandra* X-ray Telescope we have detected several local X-ray absorption lines along lines-of-sight towards distant quasars. These absorption lines are produced by warm/hot gas located in local intergalactic space and/or in our Galaxy. I will present our observations and discuss the origin of the X-ray absorption and its implications in probing the warm/hot component of local baryons.

1. Introduction

The cosmic baryon budget at low and high redshift indicates that a large fraction of baryons in the local universe have so far escaped detection (e.g., Fukugita, Hogan, & Peebles 1998). While there is clear evidence that a significant fraction of these "missing baryons" (between 20-40% of total baryons) lie in photoionized, low-redshift Lyα clouds (Penton, Shull, & Stocke 2000), the remainder could be located in intergalactic space with temperatures of $10^5 - 10^7$ K (warm-hot intergalactic medium, or WHIM). Resonant absorption from highly-ionized ions located in the WHIM gas has been predicted based on both analytic studies of structure formation and evolution (Shapiro & Bahcall 1981;Perna & Loeb 1998;Fang & Canizares 2000) and cosmic hydrodynamic simulations (Hellsten, Gnedin, & Miralda-Escudé 1998;Cen & Ostriker 1999a;Davé et al. 2001;Fang, Bryan, & Canizares 2002). Recent discovery of O VI absorption lines by the Hubble Space Telescope (*HST*) and the Far Ultraviolet Spectroscopic Explorer (*FUSE*) (see, e.g., Tripp & Savage 2000) indicates that there may be a significant reservoir of baryons in O VI absorbers. While Li-like O VI probes about $\sim 30 - 40\%$ of the WHIM gas (Cen et al. 2001;Fang & Bryan 2001), the remaining $\sim 60 - 70\%$ is hotter and

R. Lieu and J. Mittaz (eds.), Soft X-Ray Emission from Clusters of Galaxies and Related Phenomena, 65–70.

can only be probed by ions with higher ionization potentials, such as H- and He-like Oxygen, through X-ray observation.

Recently, with *Chandra* Low Energy Transmission Grating Spectrometer (LETGS) we detected resonance absorption lines from H- and He-like Oxygen in the X-ray spectra of background quasars, namely PKS 2155-304 and 3C 273. The detected lines can be categorized into (1) those at $z \approx 0$ and (2) one redshifted intervening system. In this paper, we will discuss these detections and their implications for the physical properties of the hot gases that give rise to these absorption features.

Table 1: Fitting parameters of the X-ray absorption Lines

	PKS 2155-304		3C 273
	O VIII Lyα	O VII Heα	O VII Heα
λ_{obs}	$20.02^{+0.015}_{-0.015}$	$21.61^{+0.01}_{-0.01}$	$21.60^{+0.01}_{-0.01}$
$cz(\mathrm{km\,s^{-1}})$	16624 ± 237	112^{+140}_{-138}	-26^{+140}_{-140}
Line Width[a]	< 0.039	< 0.027	< 0.020
Line Flux[b]	$4.8^{+2.5}_{-1.9}$	$5.5^{+3.0}_{-1.7}$	$4.2^{+1.8}_{-0.9}$
EW (mÅ)	$14.0^{+7.3}_{-5.6}$	$15.6^{+8.6}_{-4.9}$	$28.4^{+12.5}_{-6.2}$
SNR	4.5	4.6	6.4

a. 90% upper limit of the line width σ, in units of Å.
b. Absorbed line flux in units of 10^{-5} photons cm^{-2}s^{-1}.

2. Data Reduction

PKS 2155-304 and 3C 273 are bright extragalactic X-ray sources used as *Chandra* calibration targets. They were observed with the *Chandra* LETG-ACIS (the observations ids for PKS 2155 are 1703, 2335, 3168; and the ids for 3C 273 are 1198, 2464, 2471). For detailed data analysis, we refer to Fang et al. 2002. We found all continua are well described by a single power law absorbed by Galactic neutral hydrogen.

After a blind search for any statistically significant absorption features, several absorption features with S/N > 4 were detected in the spectra of both quasars in the 2–42 Å region of the LETGS spectral bandpass (Figure 1). These features were subsequently fit in ISIS (Houck & Denicola 2000).

3. Discussion

3.1 PKS 2155-304

The absorption feature at ~ 21.6 Å was reported by Nicastro et al. (2002) in the LETGS-HRC archival data. We concentrate on the absorption feature which appears at 20.02 Å (619 eV). Considering cosmic abundances and oscillator strengths for different ions, O VIII Lyα is the only strong candidate line between 18 and 20 Å, the measured wavelength de-redshifted to the source. It is plausible that the 20 Å absorption is due to O VIII Lyα in a known intervening system at $cz \approx 16,734$ km s^{-1}. With *HST*, Shull et al. (1998) discovered a cluster of low metallicity H I Lyα clouds along the line-of-sight (LOS) towards PKS 2155-304, most of which have redshifts between $cz = 16,100$ km s^{-1} and $18,500$ km s^{-1}. Using 21 cm images from the *Very Large Array* (*VLA*), they detected a small group of four H I galaxies offset by $\sim 400 - 800$ h$_{70}^{-1}$ kpc from the LOS, and suggested that the H I Lyα clouds could arise from gas associated with the group (We use $H_0 = 70h_{70}$ km s^{-1}Mpc^{-1} throughout the paper).

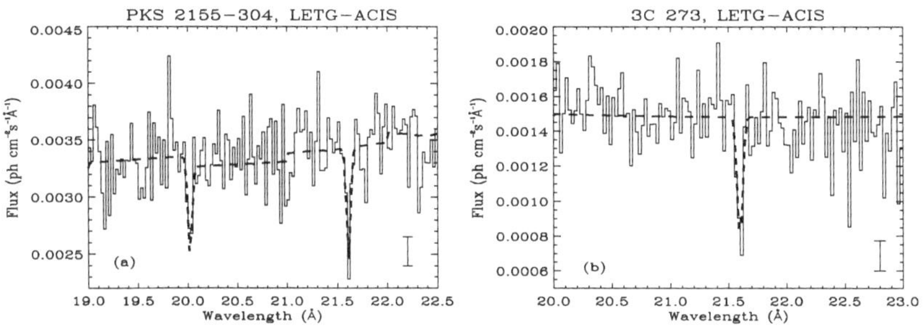

Figure 1. The *Chandra* LETG-ACIS spectra of (a) PKS 2155 and (b) 3C273. The dashed lines are the fitted spectra. The average 1σ error bar plotted on the right-bottom of each panel is based on statistics only.

Taking the absorption line to be O VIII Lyα, we estimate the column density is N(O VIII) $\sim 9.5 \times 10^{15}$ cm^{-2} if the line is unsaturated. We can constrain the density of the absorbing gas, assuming it is associated with the intervening galaxy group. Since the line is unresolved, a lower limit of $n_b > (1.0 \times 10^{-5}$ cm$^{-3})$ $Z_{0.1}^{-1} f_{0.5}^{-1} l_8^{-1}$ can be obtained. Here $Z_{0.1}$ is the metallicity in units of 0.1 solar abundance, $f_{0.5}$ is the ionization fraction in units of 0.5, and l_8 is the path length in units of $8h_{70}^{-1}$Mpc. A more reasonable estimate of the path length comes from the mean projected separation of $\sim$1 Mpc for the galaxies in the group, which gives $n_b \approx 7.5 \times 10^{-5}$ cm^{-3} $Z_{0.1}^{-1} f_{0.5}^{-1}$. This implies a range of baryon overdensity ($\delta_b \sim$ 50 - 350) over the cosmic mean $\langle n_b \rangle = 2.14 \times 10^{-7}$ cm^{-3}. Interestingly, Shull et al.(1998) estimate an overdensity for the galaxy group of $\delta_{gal} \sim$100.

In the case of pure collisional ionization, temperature is the only parameter of importance over a wide range of density so long as the gas is optically thin. The O VIII ionization fraction peaks at 0.5, and exceeds 0.1 for temperatures $T \sim 2 - 5 \times 10^6$ K. Using CLOUDY (Ferland et al. 1998) we find that photoionzation by the cosmic UV/X-ray background is not important for $n_b > 10^{-5}$ cm^{-3}. CLOUDY calculations of the column density ratios between other ions and O VIII also show that $T \gtrsim 10^{6.4}$ K.

Assuming conservative upper limits of $Z \lesssim 0.5Z_\odot$ and an O VIII ionization fraction of $f \lesssim 0.5$, and a path length of $\Delta z \lesssim 0.116$, we estimate $\Omega_b(\mathrm{O\ VIII}) \gtrsim 0.005h_{70}^{-1}$. This is about 10% of the total baryon fraction, or about 30-40% of the WHIM gas, if the WHIM gas contains about 30-40% of total baryonic matter. This baryon fraction is consistent with the prediction from Perna & Loeb (1998) based on a simple analytic model.

3.2 3C 273

Based on the detected line equivalent width (W_λ) and non-detection of local O VII Heβ line at $18.6288\mathring{A}$, we estimate the column density $\mathrm{N(O\ VII)} = 1.8^{+0.2}_{-0.7} \times 10^{16}\ \mathrm{cm}^{-2}$ if the line is unsaturated[1]. The O VII indicates the existence of the gas with high temperatures $\sim 10^6$ K. The extremely high temperatures imply that O VII is unlikely to to be produced in the nearby interstellar medium (ISM), although we cannot rule out the possibility of an origin from supernova remnants. It is more plausible that the He-like Oxygen is produced in distant, hot halo gas, or even in the Local Group (LG).

3.2.1 Local Group Origin?

We can constrain the density of the absorbing gas, assuming it is associated with the Local Group. The path length can be set equal to the distance to the boundary of the Local Group, where the gas begins to participate in the Hubble flow. Assuming a simple geometry for the Local Group, the path length is ~ 1 Mpc (see the following text). Adopting the 90% lower limit of the O VII column density, this gives $n_b > (2.2 \times 10^{-5}\ \mathrm{cm}^{-3})\ Z_{0.2}^{-1} f_1^{-1} l_1^{-1}$, where $Z_{0.2}$ is the metallicity in units of 0.2 solar abundance, f_1 is the ionization fraction, and l_1 is the path length in units of 1 Mpc.

Assuming spherical symmetry and isothermality, a model characterizing the distribution of the Local Group gas is given by the standard β-model. We adopt a simplified geometry model of the LG, where the LG barycenter is located along the line connecting M31 and the Milky Way, at about 450 kpc away from our Galaxy (Rasmussen & Pedersen 2001). At about $r \sim 1200$ kpc the gravitational contraction of the Local Group starts to dominate the Hubble flow, and this was defined as the boundary of the Local Group (Courteau & van den Bergh 1999). Based on this simple model we estimate the column density of O VII by integrating the O VII number density $n_{O\ VII}$ along this path length. We find a tight upper limit of Local Group temperature $\mathrm{T} \leq 1.2 \times 10^6$ K; at temperatures higher than 1.2×10^6 K, the O VII ionization fraction drops quickly and O VIII starts to dominate. We also find that the temperature of the Local Group should be higher than 2.3×10^5 K. To satisfy the observed O VII column density we find that the gas distribution should have a rather flat core with $r_c \geq 100$ kpc.

3.2.2 Hot Halo Gas?

Strong O VI absorption ($\log N(\mathrm{O\ VI}) = 14.73 \pm 0.04$) along the sight line towards 3C 273 was detected with *FUSE* between -100 and +100 $\mathrm{km\ s^{-1}}$ (Sembach et al. 2001). This absorption probably occurs in the interstellar medium of the Milky Way disk and halo. Absorption features of lower ionization species are also present at these velocities. O VI absorption is also detected between +100 and +240 $\mathrm{km\ s^{-1}}$ in the form of a broad, shallow absorption wing extending redward of the primary Galactic absorption feature with $\log N(\mathrm{O\ VI}) = 13.71$. The O VI absorption wing has been attributed to hot gas flowing out of the Galactic disk as part of a "Galactic chimney" or "fountain" in the Loop IV and North Polar Spur regions of the sky. Alternatively, the wing might be remnant tidal debris from interactions of the Milky Way and smaller Local Group galaxies (Sembach et al. 2001). It is possible to associate the *Chandra*-detected O VII absorption with these highly ionized metals detected by *FUSE*. Here we discuss several scenarios:

(1). The O VII is related to the primary O VI feature: In this case, $\log N(\mathrm{OVI}) = 14.73$, and $\log[N(\mathrm{O\ VII})/N(\mathrm{O\ VI})] \sim 1.5$, assuming $N(\mathrm{O\ VII}) = 1.8 \times 10^{16}\ \mathrm{cm^{-2}}$. This is within about a factor of 2 of the O VII/O VI ratio observed for the PKS 2155-304 absorber and is consistent with the idea that the gas is radiatively cooling from a high temperature (Heckman et al. 2002). This possibility is appealing since the centroids of the O VI and O VII absorption features are similar ($\sim 6 \pm 10\ \mathrm{km\ s^{-1}}$ versus $-26 \pm 140\ \mathrm{km\ s^{-1}}$), and the width of the resolved O VI line (FWHM $\sim 100\ \mathrm{km\ s^{-1}}$) is consistent with a broad O VII feature. However, this possibility also has drawbacks that the predicted O VIII column density is too high and the amount of C IV predicted is too low.

(2). The O VII is related to the O VI wing: In this case, the O VII is associated only with the O VI absorption "wing". This seems like a reasonable possibility. Then $\log[N(\mathrm{O\ VII})/N(\mathrm{O\ VI})] \sim 2.5$. In collisional ionization equilibrium, this would imply a temperature of $> 10^6$ K (Sutherland & Dopita 1993). The non-detection of O VIII Lyα absorption requires the temperatures lower than $\sim 10^{6.3}$ K.

(3). The O VII is related to none of the O VI: In this case, the temperature should be high enough to prevent the production of O VI ions. We can reach the similar conclusions to those in situation (2).

We thank members of the MIT/CXC team for their support. This work is supported in part by contracts NAS 8-38249 and SAO SV1-61010. KRS acknowledges financial support through NASA contract NAS5-32985 and Long Term Space Astrophysics grant NAG5-3485.

Notes

1. Rasmussen et al.(2002) also reported the detection of a similar feature with *XMM*-Newton in this conference.

References

Cen, R. & Ostriker, J.P. 1999a, ApJ, 514, 1
Cen, R. & Ostriker, J. P. 1999b, ApJ, 519, L109
Cen, R., Tripp, T.M., Ostriker, J.P. & Jenkins, E.B. 2001, ApJ, 559, L5
Courteau, S. & van den Bergh, S. 1999, AJ, 118, 337
Davé, R. et al. 2001, ApJ, 552, 473
Fang, T. & Bryan, G. L. 2001, ApJ, 561, L31
Fang, T., Bryan, G.L. & Canizares, C.R. 2002, ApJ, 564, 604
Fang, T. & Canizares, C. R. 2000, ApJ, 539, 532
Fang, T. et al. 2002, ApJ, 572, L127
Ferland, G.J. et al. 1998, PASP, 110, 761
Fukugita, M., Hogan, C. J., & Peebles, P. J. E. 1998, ApJ, 503, 518.
Houck, J. C. & Denicola, L. A. 2000, ASP Conf. Ser. 216: Astronomical Data Analysis Software and Systems IX, 9, 591
Heckman, T. M. et al 2002, ApJ, submitted (astro-ph/0205556)
Hellsten, U., Gnedin, N. Y., & Miralda-Escudé, J. 1998, ApJ, 509, 56
Nicastro, F. et al. 2002, ApJ, 573, 157
Penton, S. V., Shull, J. M., & Stocke, J. T. 2000, ApJ, 544, 150.
Perna, P. & Loeb, A. 1998, ApJ,503, L135
Rasmussen, J. & Pedersen, K. 2001, ApJ, 559, 892
Rasmussen, A. et al. 2002, this proceeding
Sembach, K. R. et al. 2001, ApJ, 561, 573
Shapiro, P. R. & Bahcall, J. N. 1981, ApJ, 245, 335
Shull, J. M. et al. 1998, AJ, 116, 2094
Sutherland, R. S. & Dopita, M. A. 1993, ApJS, 88, 253
Tripp, T.M. & Savage, B.D. 2000, ApJ, 542, 42

THE FUSE SURVEY OF O VI IN THE GALACTIC HALO

B.D. Savage[1], K.R. Sembach [2], B.P. Wakker[1], P. Richter[3], M. Meade[1], E.B Jenkins[4], J.M. Shull[5], H.W. Moos[6], and G. Sonneborn[7]

[1]*Department of Astronomy, University of Wisconsin, Madison, WI*

[2]*Space Telescope Science Institute, Baltimore, MD*

[3]*Osservatoria Astrofisico di Arcetri, Firenzi, Italy*

[4]*Princeton University Observatory, Peyton Hall, Princeton, NJ*

[5]*Center for Astrophysics and Space Astronomy, University of Colorado, Boulder, CO*

[6]*Department of Physics and Astronomy, Johns Hopkins University, Baltimore, MD*

[7]*Laboratory for Astronomy and Solar Physics, NASA, Greenbelt, MD*

Abstract We summarize the results of the Far-Ultraviolet Spectroscopic Explorer (FUSE) program to study O VI in the Milky Way halo. Spectra of 100 extragalactic objects and two distant halo stars are analyzed to obtain measures of O VI absorption along paths through the Milky Way thick disk/halo. Strong O VI absorption over the velocity range from -100 to 100 km s^{-1} reveals a widespread but highly irregular distribution of O VI, implying the existence of substantial amounts of hot gas with $T \sim 3 \times 10^5$ K in the Milky Way thick disk/halo. The overall distribution of O VI can be described by a plane-parallel patchy absorbing layer with an average O VI mid-plane density of n_o(O VI) = 1.7×10^{-8} cm^{-3}, an exponential scale height of $\sim$2.3 kpc, and a $\sim$0.25 dex excess of O VI in the northern Galactic polar region. The distribution of O VI over the sky is poorly correlated with other tracers of gas in the halo, including low and intermediate velocity H I, Hα emission from the warm ionized gas at $\sim 10^4$ K, and hot X-ray emitting gas at $\sim 10^6$ K. The O VI has an average velocity dispersion, b=60 km s^{-1} and small standard deviation of 15 km s^{-1}. Thermal broadening alone cannot explain the large observed profile widths. A combination of models involving the radiative cooling of hot fountain gas, the cooling of supernova bubbles in the halo, and the turbulent mixing of warm and hot halo gases is required to explain the presence of O VI and other highly ionized atoms found in the halo.

R. Lieu and J. Mittaz (eds.), Soft X-Ray Emission from Clusters of Galaxies and Related Phenomena, 71–81.

1. Introduction

Absorption line observations of the highly ionized lithium-like atoms of O VI, N V, and C IV provide valuable information about gas in interstellar space with temperatures of $\sim 3\times10^5$ to 10^5 K. Among these three species, O VI is especially important because of the large cosmic abundance of oxygen and the large energy (113.9 eV) required to convert O V into O VI. Studies of interstellar O VI have been hampered by the difficulties involved in making observations in the far-UV wavelength range of its resonance doublet at 1031.93 and 1037.62 Å. Instruments operating efficiently at wavelengths shortward of $\sim$1150 Å require reflecting optics with special coatings (such as LiF or SiC) and windowless detectors. Although the Copernicus satellite successfully observed interstellar O VI toward many stars in the Galactic disk (Jenkins 1978a,b), the studies were limited to stars with visual magnitudes $m_V<7$ and did not yield information about the extension of O VI away from the Galactic plane.

Except for brief observing programs with the Hopkins Ultraviolet Telescope (HUT; Davidsen 1993) and the spectrographs in the Orbiting and Retrievable Far and Extreme Ultraviolet Spectrometers (ORFEUS; Hurwitz & Bowyer 1996; Hurwitz et al. 1998; Widmann et al. 1998; Sembach, Savage, & Hurwitz 1999), the study of O VI in the Milky Way halo has required the high throughput capabilities of the Far-Ultraviolet Spectroscopic Explorer (FUSE) satellite launched in 1999 (Moos et al. 2000; Sahnow et al. 2000).

2. Observations and Reductions

The FUSE O VI catalog paper (Wakker et al. 2003) contains the full details of the FUSE observations and reductions of Galactic O VI absorption toward 100 extragalactic sources and two halo stars. (The O VI λ1037.62 line is often confused by blending with C II* λ1037.02 and the H_2 (5-0) R(1) and P(1) lines at 1037.15 and 1038.16 Å. The O VI λ1031.93 line is usually relatively free of blending, since the H_2 (6-0) P(3) and R(4) lines are often relatively weak and are displaced in velocity by -214 and 125 km s^{-1} from the rest O VI velocity.

Information about O VI in the Milky Way halo (see Savage et al. 2003) has mostly been derived from the relatively blend-free O VI λ1031.93 line. Figure 1 shows O VI λ1031.93 line profiles versus LSR velocity for a sample of 12 objects. The continuum placement is generally quite reliable since most of the objects are AGNs, which usually have well-defined power-law continua. Along sight lines where the H_2 (6-0) P(3) and R(4) lines at 1031.19 Å and 1032.35 Å blend with O VI, the expected strength of the contamination is based on an analysis of other H_2 J=3 and 4 absorption lines in the spectrum. The estimated H_2 absorption line profiles have been included in Figure 1. For O VI at 3×10^5 K (the temperature at which O VI peaks in abundance) the ther-

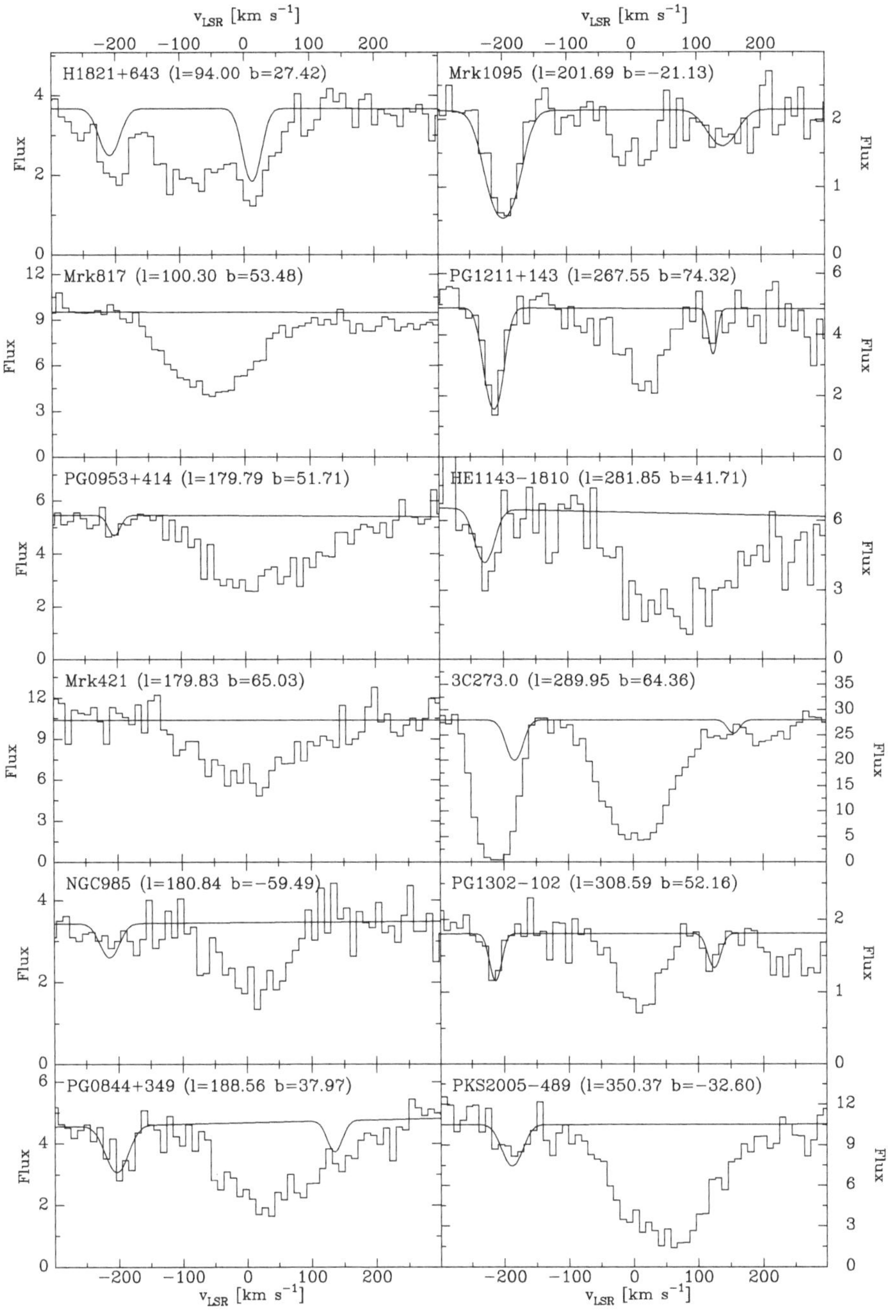

Figure 1. A sample of the O VI λ1031.93 absorption line profiles from Wakker et al. (2003) is displayed. Flux (10^{-14} erg cm^{-2} s^{-1} Å^{-1}) is plotted against LSR velocity with the solid line showing the continuum placement including an estimate of the contaminating H_2 (6-0) R(3) and R(4) absorption when necessary (from Savage et al. 2003).

mal Doppler contribution to the O VI line width corresponds to b=17.7 km s^{-1} (FWHM=29.4 km s^{-1}), where b is the standard Doppler spread parameter. The expected thermal width of the O VI line is comparable to the FUSE resolution of 20–25 km s^{-1}. Since the O VI lines are usually resolved, reliable column densities can be obtained using the apparent optical depth method (Savage & Sembach 1991). The O VI column densities, velocities and line widths can be found in Savage et al. (2003).

3. Separating Low Velocity Thick O VI Disk Absorption from High Velocity Absorption

The lines of sight to the objects observed in the FUSE halo gas O VI program pass through Milky Way disk gas, thick disk/halo gas, intermediate-velocity clouds (IVCs), high-velocity clouds (HVCs), and may even sample intergalactic O VI in the Local Group of galaxies. At high Galactic latitude the dividing lines in velocity between high, intermediate, and low velocity are generally at $|v_{LSR}|$ = 90 and 30 km s^{-1}, although the effects of Galactic rotation must also be considered. The O VI absorption lines trace a complex set of processes and phenomena involving hot gas in the Milky Way disk, thick disk/halo, and beyond. We refer to O VI in the Milky Way disk-halo interface extending several kpc away from the Galactic plane as the "thick disk O VI".

The O VI profiles exhibit a diversity of strengths and kinematical behavior. Disk, and thick disk O VI are clearly detected within the general velocity range -90 to 90 km s^{-1} toward 91 of the 102 survey objects. In addition, the absorption profiles reveal O VI at high velocities with $|v_{LSR}|$ ranging from $\sim$100 to 400 km s^{-1} in $\sim$60% of the sight lines. The O VI absorption at high-velocity is considered by Sembach et al. (2003).

4. Column Density Distribution for the Thick Disk Absorption

The relatively strong O VI absorption associated with the thick disk of Galactic O VI has log N(O VI) between 13.85 and 14.78. The equivalent column density perpendicular to the Galactic plane, log [N(O VI)sin$|b|$], ranges from 13.45 to 14.68. Figures 2a and 2b show histograms of the distribution of log N(O VI) and log [N(O VI)sin$|b|$]. The distribution of O VI on the sky is quite irregular. In Figure 3 the O VI measurements are superposed on a grey scale representation of the 0.25 keV X-ray background as recorded by ROSAT (Snowden et al. 1997).

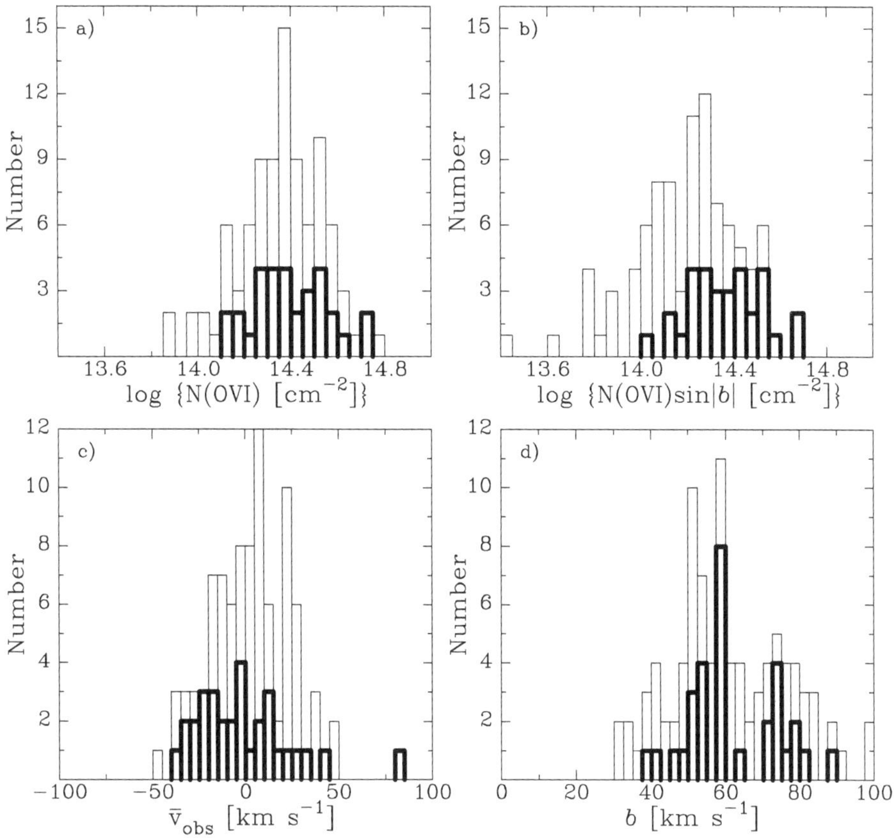

Figure 2. Number distribution of log N(O VI), log [N(O VI)sin|b|], v_{obs} (km s^{-1}), and b (km s^{-1}) for O VI absorption associated with the thick disk of the Milky Way are shown in (a), (b), (c), and (d), respectively. In the histograms for log N(O VI) and log [N(O VI)sin|b|] upper limits are indicated as detections. In the various panels the upper histogram is for the full object sample while the heavy line histogram is for survey objects with |b|>45° (from Savage et al. 2003).

5. The Extension of O VI Away from the Galactic Plane

The large O VI column densities measured along extragalactic sight lines imply that a substantial amount of O VI is situated in the Galactic thick disk at large distances away from the plane of the Galaxy. Figure 4 shows log [N(O VI)sin|b|] versus log |z(kpc)| for various data samples, as detailed in the figure caption. There is a considerable spread in the values of log [N(O VI)]sin|b|] at all values of |z|. The data points for the LMC and SMC directions at |z|=27 and 50 kpc, respectively, show the spread in values of log[N(O VI) sin|b|] over the relatively small angular extents of these two galaxies. In contrast the spread for the Copernicus observations and the extragalactic observa-

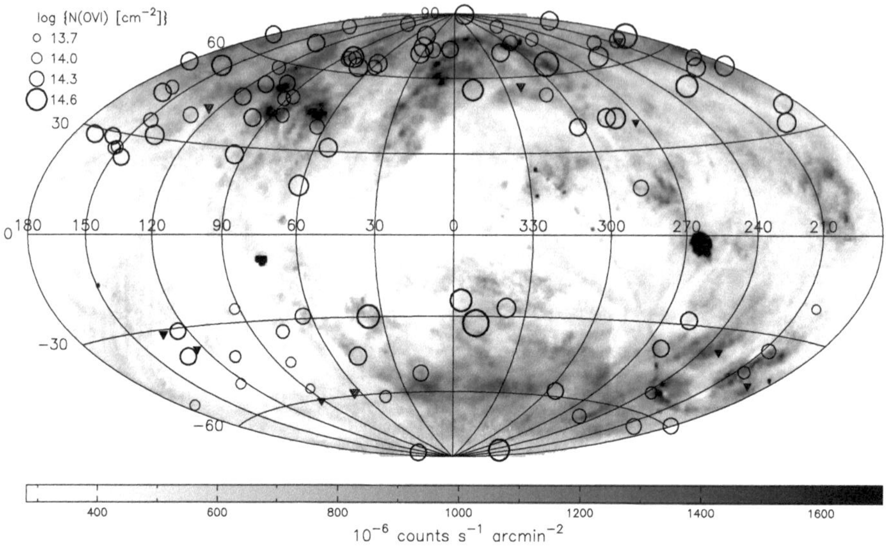

Figure 3. Values of log N(O VI) for the thick disk of the Milky Way are represented as circles displayed in these aitoff projections of the sky with the Galactic center at the center of the figures and Galactic longitude increasing to the left. The circle size is proportional to log N(O VI) according to the code shown. Upper limits to log N(O VI) are denoted with triangles with a size proportional to the limit. The grey scale shows the 0.25 keV X-ray sky diffuse background count rate as measured by the ROSAT Satellite (Snowden et al. 1997) (from Savage et al. 2003).

tions illustrate the variation in the values of log [N(O VI)sin$|b|$] extending over most of the sky.

The three solid lines in Figure 4 show the expected behavior of the log [N(O VI)sin$|b|$] versus log $|z(\mathrm{kpc})|$ distribution for a smoothly distributed, exponentially stratified plane-parallel Galactic atmosphere with an O VI mid-plane density n_o(O VI)= 1.7×10^{-8} cm^{-3} and exponential scale heights of 1.0, 2.5, and 10 kpc. The use of a mid-plane density of 1.7×10^{-8} cm^{-3} appears well justified since the value represents an average extending over ~220 stars in the combined Copernicus and FUSE disk star sample (Jenkins et al. 2001). The large irregularity in the distribution of O VI and the enhancement in the amount of O VI over the northern Galactic hemisphere introduces a complication when trying to estimate a Galactic scale height for O VI.

The data points in Figure 4 are consistent with an O VI scale height in the range between 1.0 and 10 kpc. The observations are not well fitted by a symmetric plane parallel model for the distribution of O VI extending away from the Galactic plane. One can imagine a wide range of possible models to fit the observed values of log [N(O VI)sin$|b|$] that are more complicated. The simplest model that improves the fit is a superposition of a plane-parallel patchy

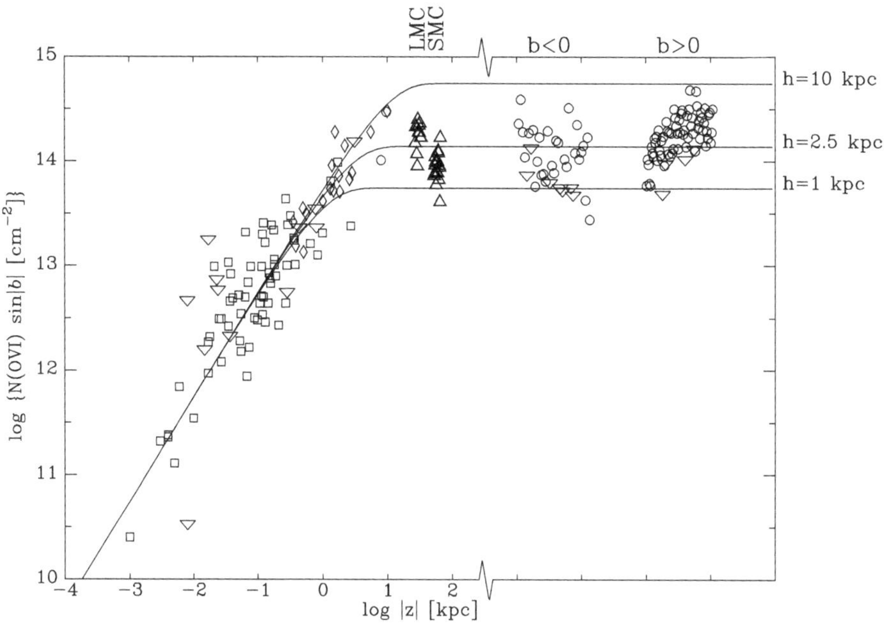

Figure 4. log [N(O VI)sin$|b|$] for gas in the Milky Way disk and thick disk is plotted against log $|z$(kpc)$|$ for (1) stars from the Copernicus sample of stars in the Galactic disk (Jenkins 1978a, open squares); (2) the FUSE 22 halo star survey (Zsargo et al. 2003, open diamonds); (3) Milky Way absorption toward stars in the LMC and SMC (Howk et al. 2002 and Hoopes et al. 2002, open upward pointing triangles); and (4) the 100 extragalactic and two Galactic stellar lines of sight analyzed by us (Savage et al. 2003, open circles, with the 100 extragalactic objects plotted on the right hand side of the figure in the region beyond the break in the log $|z$(kpc)$|$ axis and the two stars plotted near log $|z|\sim$1.0; 3σ upper limits are plotted as the downward pointing triangles). The three solid lines show the expected behavior of the log N(O VI)sin$|b|$ versus log $|z$(kpc)$|$ distribution for a smoothly distributed exponentially stratified plane parallel Galactic atmosphere with an O VI mid-plane density n_o(O VI)=1.7$\times$10^{-8} cm^{-3} and exponential scale heights 1.0, 2.5 and 10 kpc (adapted from Savage et al. 2003).

absorbing layer with an exponential scale height of $\sim$2.3 kpc, a mid-plane density of 1.7$\times$10^{-8} cm^{-3}, and a $\sim$0.25 dex excess of O VI at the higher northern Galactic latitudes.

6. Kinematics of the Thick Disk O VI

The kinematic properties of the O VI absorption provide useful information about the various physical processes controlling the distribution of gas at $T\sim3\times10^5$ K in the Galactic halo. Two simple measures of the kinematics of the O VI absorption studied by Savage et al. (2003) are the average LSR velocity, and the velocity dispersion of the absorption as measured by the Doppler-spread parameter b. The distributions of the values of $v_{\rm obs}$ and b are shown in Figures 2c and 2d. In these figures the upper (light line) histogram is for the complete sample, while the lower (heavy line) histogram is for high latitude objects ($|b|>45°$). At high latitudes, where the effects of Galactic rotation are small, the values of v_{obs} exhibit a large spread in velocity with an average of 0 km s^{-1} and a standard deviation of 22 km s^{-1}. It is interesting that the O VI at high latitudes in each Galactic hemisphere is moving with positive and negative velocity with nearly equal frequency. For the full sample, b(min)=30 km s^{-1}, b(max)=99 km s^{-1}, b(median)=59 km s^{-1}, b(average)=61 km s^{-1}, and a the standard deviation on b is 15 km s^{-1}. The O VI absorption lines are much wider than the absorption expected from thermal Doppler broadening alone since gas at $T{\sim}3{\times}10^5$ K, the temperature at which O VI is expected to peak in abundance, has b(O VI)=17.7 km s^{-1}. Inflow, outflow, Galactic rotation, and turbulence therefore also affect the profiles.

7. O VI Versus Other ISM Tracers

The relationship between the column density of O VI and other ISM tracers was studied for H I, Hα, soft X-ray emission, and non-thermal radio emission. In all cases the amount of O VI is poorly related to the amount of these other ISM tracers. One example is shown in Figure 3 where log N(O VI) is compared to the 0.25 keV soft X-ray background count rate from Snowden et al. (1997) at a resolution of 36′. The correspondence between the X-ray sky brightness and N(O VI) is poor. This is not too surprising given the 0.25 Kev diffuse X-ray background is a complex superposition of a non-uniform local bubble component and halo and extragalactic background components (Snowden et al. 1998, 2000) experiencing attenuation due to photoelectric absorption occurring in cooler foreground gas.

8. Origin of O VI and Other Highly Ionized Species in the Galactic Halo

The FUSE O VI survey observations reported here provide important new insights into the distribution and kinematics of highly ionized gas in the Milky Way halo. The observations confirm the basic validity of Spitzer's (1956) prediction that the ISM of the Galaxy contains a hot gas phase that extends well away from the Galactic plane. Theories for the origin of highly ionized gas in the Milky Way halo must explain the distribution, ionization, kinematics, and support of the gas. For reviews of the models see Spitzer (1990), McKee (1993), and Savage (1995). The ionization of Si IV and C IV is likely either from electron collisions in a hot gas, photoionization, or some combination of both processes. With their high ionization threshold, O VI and N V are more likely to be produced by collisional ionization in hot gas. However, non-equilibrium ionization effects will probably be important because collisionally ionized gas cools very rapidly in the temperature range (1–5)$\times 10^5$ K. Such transition temperature gas might occur within the cooling gas of a "Galactic fountain" (Shapiro & Field 1976; Edgar & Chevalier 1986; Shapiro & Benjamin 1991), in the conductively heated interface region between the hot and cool interstellar gas (Ballet, Arnaud, & Rothenflug 1986), in radiatively cooling SN bubbles (Slavin & Cox 1992, 1993) or in turbulent mixing layers (TMLs) where hot gas and warm gas are mixed by turbulence to produce gas with non-equilibrium ionization characteristics (Begelman & Fabian 1990; Slavin, Shull, & Begelman 1993). The heating and ionization of the gas could also occur through magnetic reconnection processes (Raymond 1992; Zimmer, Lesch, & Birk 1997). With a number of processes probably contributing to the O VI found in the halo, it will be difficult to clearly identify the most important processes.

9. Summary

FUSE far-UV spectra of 100 extragalactic objects and two halo stars are used to obtain measures of O VI far-UV absorption along paths through the Milky Way halo. High velocity O VI absorption with $|v|>100$ km s^{-1} is seen along $\sim$60% of the sight lines (see Sembach et al. 2003). In this paper we review the results found in the study the Milky Way thick-disk O VI absorption that covers the approximate velocity range from -100 to 100 km s^{-1} (see Savage et al. 2003). The results are:

1. There exists a widespread but highly irregular distribution of O VI in the Galactic thick disk, implying the existence of substantial amounts of hot gas with $T\sim 3\times 10^5$ K. The integrated O VI logarithmic column density through the halo, log N(O VI), ranges from 13.85 to 14.78 with an average of 14.38.

2. Averages of log [N(O VI)sin$|b|$] over Galactic latitude reveal a $\sim$0.25 dex excess of O VI to the north Galactic polar region compared to the rest of the Galaxy.

3. The observations are not well described by a simple symmetrical plane-parallel patchy distribution of O VI absorbing structures. The simplest departure from such a model that provides a better fit to the observations is a plane-parallel patchy absorbing layer with an average O VI mid-plane density of n_o(O VI)=1.7×10^{-8} atoms cm^{-3} and a scale height of $\sim$2.3 kpc combined with a $\sim$0.25 dex excess of O VI absorbing gas in the northern Galactic polar region.

4. The O VI is poorly correlated with other ISM tracers of gas in the halo, including low and intermediate velocity H I, Hα emission from the warm ionized medium, and hot 0.25 keV X-ray emitting gas.

5. The O VI profiles range in velocity dispersion, b, from 30 to 99 km s^{-1}, with an average value of 61 km s^{-1} and standard deviation of 15 km s^{-1}.

The thermal Doppler component of the broadening cannot explain the large observed profile widths since gas at $T{\sim}3{\times}10^5$ K, the temperature at which O VI is expected to peak in abundance in collisional ionization equilibrium, has b(O VI)=17.7 km s^{-1}. Inflow, outflow, Galactic rotation, and turbulence therefore also affect the profiles.

6. The O VI average absorption velocities for thick disk gas toward high latitude objects ($|b|{>}45^{\circ}$) range from -37 to 82 km s^{-1}, with a high latitude sample average of 0 km s^{-1} and a standard deviation of 21 km s^{-1}. Thick disk O VI is observed to be moving both toward and away from the plane with roughly equal frequency.

7. The broad high positive velocity O VI absorption wings extending from $\sim$100 to $\sim$250 km s^{-1} seen in the spectra of 21 objects may be tracing the outflow of gas into the halo, although we can not rule out a more distant origin.

8. A combination of models involving the radiative cooling of hot gas in a Galactic fountain flow, the cooling of hot gas in halo supernova bubbles, and the turbulent mixing of warm and hot halo gases appears to be required to explain the highly ionized atoms found in the halo. If the origin of the O VI is dominated by a fountain flow, a mass flow rate of approximately 1.4 $M_{\odot}$ yr^{-1} to each side of the Galactic disk for cooling hot gas with an average density of 10^{-3} cm^{-3} is required to explain the average value of log [N(O VI)sin$|b|$ found in the southern Galactic hemisphere.

This work is based on data obtained for the Guaranteed Time Team by NASA-CNES-CSA FUSE mission operated by Johns Hopkins University. Financial support to U.S. participants has been provided by NASA contract NAS5-32985. K.R.S. acknowledges additional financial support through NASA Long Term Space Astrophysics grant NAG5-3485. B.P.W. acknowledges additional support from NASA grants NAG5-9179, NAG5-9024, and NGG5-8967.

References

[sv1]sv1Ballet, J., Arnaud, M., & Rothenflug, R. 1986, A&A, 161, 12

Begelman, M.C., & Fabian, A.C. 1990, MNRAS, 244, 26p

Davidsen A.F. 1993, Science, 259, 327

Edgar, R.J., & Chevalier, R.A. 1986, ApJ, 310, L27

Hoopes, C.G., Sembach, K.R., Howk, J.C., Savage, B.D., & Fullerton, A.W. 2002, ApJ, 569, 233

Howk, J.C., Savage, B.D., Sembach, K.R., & Hoopes, C.G. 2002, ApJ, 572, 264

Hurwitz, M., & Bowyer, S. 1996, ApJ, 465, 296

Hurwitz, M. et al. 1998, ApJ, 500, L61

Jenkins, E.B. 1978a, ApJ, 219, 845

Jenkins, E.B. 1978b, ApJ, 220, 107

Jenkins, E.B., Bowen, D.V., & Sembach, K.R. 2001, in the Proceedings of the XVIIth IAP Colloquium: "Gaseous Matter in Galactic and Intergalactic Space", eds. R. Ferlet, M. Lemoine, J.M. Desert, B. Raban, (Frontier Group), 99

McKee, C. 1993, in "Back to the Galaxy", eds. S.G. Holt & F. Verter (New York: AIP), 499

Moos, H.W. et al. 2000, ApJ, 538, L1

Raymond, J.C. 1992, ApJ, 384, 502

Sahnow, D. et al. 2000, ApJ, 538, L7

Savage, B.D. 1995, in "The Physics of the Interstellar and Intergalactic Medium", eds. A.Ferrara, C.F.McKee, C.Heiles, & P.R. Shapiro (San Francisco: ASP Conf.Pub.) Vol.80, 233

Savage, B.D., & Sembach, K.R. 1991, ApJ, 379, 245

Savage, B.D., Sembach, K.R., Wakker, B.P., Richter, P., Meade, M., Shull, J.M., Jenkins, E.B., Sonneborn, G., & Moos, H.W. 2003, ApJS, May Issue.

Sembach, K.R., Wakker, B.P., Savage, B.D., Richter, P., Meade, M., Shull, J.M., Jenkins, E.B., Sonneborn, G., & Moos, H.W. 2003, ApJS, May Issue.

Sembach, K.R., Savage, B.D., & Hurwitz, M. 1999, ApJ, 524, 98

Shapiro, P.R., & Benjamin, R.A. 1991, PASP, 103, 923

Shapiro, P.R., & Field, G.B. 1976, ApJ, 205, 762

Slavin, J.D., & Cox, D.P. 1992, ApJ, 392, 131

Slavin, J.D., & Cox, D.P. 1993, ApJ, 417, 187

Slavin, J.D., Shull, J.M., & Begelman, M.C. 1993, ApJ, 407, 83

Snowden, S.L.et al. 1997, ApJ, 485, 125

Snowden, S.L., Egger, R., Finkbeiner, D.P., Freyberg, M.J., & Plucinsky, P.P. 1998, ApJ, 493, 715

Snowden, S.L., Freyberg, M.J., Kuntz, K.D., & Sanders, W.T. 2000, ApJS, 128, 171

Spitzer, L. 1956, ApJ, 124, 20

Spitzer, L. 1990, ARA&A, 28, 71

Wakker, B.P., Savage, B.D., Sembach, K.R., Richter, P., Meade, M. et al. 2003, ApJS, May Issue

Widmann, H. et al. 1998, A&A, 338, L1

Zimmer, F., Lesch, H., & Birk, G.T. 1997, A&A, 320, 746

Zsargo, J., Sembach, K.R., Howk, J.C., & Savage, B.D. 2003, ApJ, in press

THE FUSE SURVEY OF HIGH VELOCITY O VI IN THE VICINITY OF THE MILKY WAY

K.R. Sembach[1], B.P. Wakker[2], B.D. Savage[2], P. Richter[3], M. Meade[2], J.M. Shull[4], E.B. Jenkins[5], H.W. Moos[6], and G. Sonneborn[7]

[1]*Space Telescope Science Institute, 3700 San Martin Dr., Baltimore, MD 21218*

[2]*Astronomy Dept., University of Wisconsin, 475 N.Charter St., Madison, WI 53706*

[3]*Osservatorio Astrofisico di Arcetri, Largo E. Fermi 5, Florence, Italy*

[4]*Center for Astrophysics and Space Astronomy, University of Colorado, Boulder, CO 80309*

[5]*Princeton University Observatory, Peyton Hall, Princeton, NJ 08544*

[6]*Department of Physics & Astronomy, Johns Hopkins University, Baltimore, MD 21218*

[7]*Laboratory for Astronomy and Solar Physics, NASA/GSFC, Greenbelt, MD 20771*

Abstract We describe an extensive FUSE survey of high velocity O VI absorption along $\sim$ 100 complete sight lines through the Galactic halo. The high velocity O VI traces tidal interactions with the Magellanic Clouds, accretion of gas, outflowing material from the Galactic disk, warm/hot gas interactions in a highly extended Galactic corona, and intergalactic gas in the Local Group. Approximately 60% of the sky (and perhaps as much as 85%) is covered by high velocity hot H^+ associated with the high velocity O VI. Some of the O VI may be produced at the boundaries between warm clouds and a hot, highly-extended Galactic corona or Local Group medium. A hot Galactic corona or Local Group medium and the prevalence of high velocity O VI are expected in various galaxy formation scenarios. Additional spectroscopic data in the coming years will help to determine the ionization properties of the high velocity clouds and discriminate between the multiple types of high velocity O VI features found in this study.

1. Introduction

Observational information about the highly ionized gas in the vicinity of galaxies is required for complete descriptions of galaxy formation and evolution. In this article, we outline a program we have conducted with the Far Ultraviolet Spectroscopic Explorer (FUSE) to study the hot gas in the vicinity of the Milky Way. The study is described in detail in a series of three articles devoted to probing the highly ionized oxygen (O VI) absorption along complete

R. Lieu and J. Mittaz (eds.), Soft X-Ray Emission from Clusters of Galaxies and Related Phenomena, 83–91.

paths through the Galactic halo and Local Group. The articles include a catalog of the spectra and basic observational information (Wakker et al. 2003), a study of the hot gas in the Milky Way halo (Savage et al. 2003), and an investigation of the highly ionized high velocity gas in the vicinity of the Galaxy (Sembach et al. 2003). Here, we summarize the high velocity gas results. A companion paper by Savage et al. (this volume) summarizes the Milky Way thick disk/halo results.

The O VI $\lambda\lambda 1031.926, 1037.617$ doublet lines are the best UV resonance lines to use for kinematical investigations of hot ($T \sim 10^5 - 10^6$ K) gas in the low-redshift universe. X-ray spectroscopy of the interstellar or intergalactic gas in higher ionization lines (e.g., O VII, O VIII) is possible with XMM-Newton and the Chandra X-ray Observatory for a small number of sight lines toward AGNs and QSOs, but the spectral resolution (R $\equiv \lambda/\Delta\lambda < 400$) is modest compared to that afforded by FUSE (R $\sim 15,000$). While the X-ray lines provide extremely useful information about the amount of gas at temperatures greater than 10^6 K, the interpretation of where that gas is located, or how it is related to the $10^5 - 10^6$ K gas traced by O VI, is hampered at low redshift by the complexity of the hot ISM and IGM along the sight lines observed.

2. Properties of the High Velocity O VI

We have conducted a study of the highly ionized high velocity gas in the vicinity of the Milky Way using an extensive set of FUSE data. We summarize the results for the sight lines toward 100 AGNs/QSOs and two distant halo stars in this section (see Sembach et al. 2003; Wakker et al. 2003). For the purposes of this study, gas with $|v_{LSR}| \gtrsim 100$ km s^{-1} is typically identified as "high velocity", while lower velocity gas is attributed to the Milky Way disk and halo. A sample spectrum from the survey is shown in Figure 1.

2.1 Detections of High Velocity O VI

We have identified approximately 85 individual high velocity O VI features along the 102 sight lines in our sample. A critical part of this identification process involved detailed consideration of the absorption produced by O VI and other species (primarily H_2) in the thick disk and halo of the Galaxy, as well as the absorption produced by low-redshift intergalactic absorption lines of H I and ionized metal species. Our careful process of identifying the high velocity features, and the possible complications involved in these identifications, are described by Wakker et al. (2003). We searched for absorption in a velocity range of ± 1200 km s^{-1} centered on the O VI $\lambda 1031.926$ line. With few exceptions, the high velocity O VI absorption is confined to $|v_{LSR}| \leq 400$ km s^{-1}, indicating that the O VI features observed are either associated with the Milky Way or nearby clouds within the Local Group.

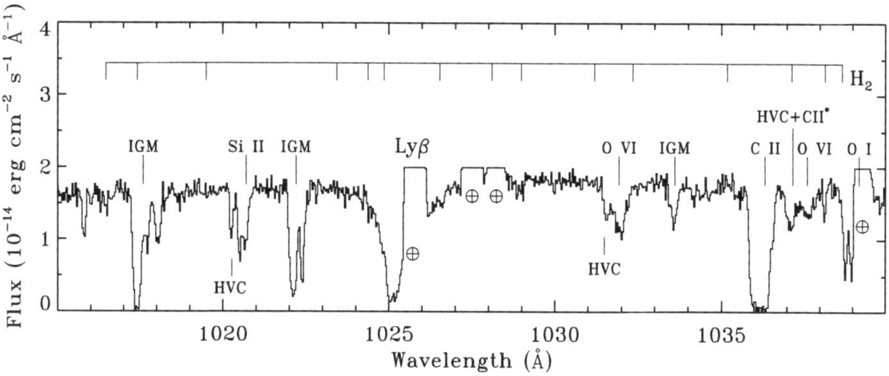

Figure 1. A portion of the FUSE spectrum of PG 1259+593 in the 1015–1040 Å spectral region. Prominent interstellar and intergalactic lines are indicated. Although weak along this sight line, the wavelengths of common H_2 lines are indicated at the top of the figure. The Si II HVC and O VI HVC absorption features identified below the spectrum trace gas in Complex C. H I and O I airglow emission lines ($\oplus$) have been truncated for clarity.

We detect high velocity O VI λ1031.926 absorption with total equivalent widths $W_\lambda > 30$ mÅ at $\geq 3\sigma$ confidence along 59 of the 102 sight lines surveyed. For the highest quality sub-sample of the dataset, the high velocity detection frequency increases to 22 of 26 sight lines. Forty of the 59 sight lines have high velocity O VI λ1031.926 absorption with $W_\lambda > 100$ mÅ, and 27 have $W_\lambda > 150$ mÅ. Converting these O VI equivalent width detection frequencies into estimates of $N(\mathrm{H}^+)$ in the hot gas indicates that $\sim 60\%$ of the sky (and perhaps as much as $\sim 85\%$) is covered by hot ionized hydrogen at a level of $N(\mathrm{H}^+) \gtrsim 8 \times 10^{17}$ cm^{-2}, assuming an ionization fraction $f_{\mathrm{O\,VI}} < 0.2$ and a gas metallicity similar to that of the Magellanic Stream ($Z \sim 0.2 - 0.3$). This detection frequency of hot H$^+$ associated with the high velocity O VI is larger than the value of $\sim 37\%$ found for high velocity warm neutral gas with N(H I)$\sim 10^{18}$ cm^{-2} traced through 21 cm emission (Lockman et al. 2002).

2.2 Velocities

The high velocity O VI features have velocity centroids ranging from $-372 < v_{LSR} < -90$ km s^{-1} to $+93 < v_{LSR} < +385$ km s^{-1}. There are an additional 6 confirmed or very likely ($> 90\%$ confidence) detections and 2 tentative detections of O VI between $v_{LSR} = +500$ and +1200 km s^{-1}; these very high velocity features probably trace intergalactic gas beyond the Local Group. Most of the high velocity O VI features have velocities incompatible with those of Galactic rotation (by definition). The dispersion about the mean of the high velocity O VI centroids decreases when the velocities are converted from the

Local Standard of Rest (LSR) into the Galactic Standard of Rest (GSR) and the Local Group Standard of Rest (LGSR) reference frames. While this reduction is expected if the O VI is associated with gas in a highly extended Galactic corona or in the Local Group, it *does not* provide sufficient proof by itself of an extragalactic location for the high velocity gas. Additional information, such as the gas metallicity or ionization state, is needed to constrain the cloud locations.

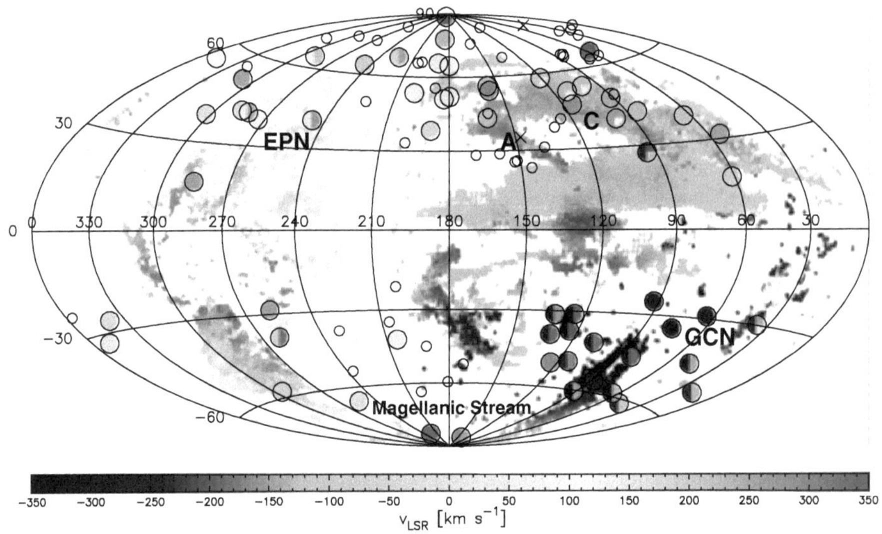

Figure 2. All-sky Hammer-Aitoff projection of the high velocity O VI features (circles) superimposed on the high velocity H I sky observed in 21 cm emission. The velocity color coding is the same for both species. Small circles indicate directions where no high velocity O VI was detected. (From Sembach et al. 2003 and Wakker et al. 2003.)

We plot the locations of the high velocity O VI features in Figure 2 with filled colored circles indicating the velocities of the O VI (blue/green = negative velocities, orange/red = positive velocities). Small open circles (or "X" marks for the two stellar sight lines) denote directions where no high velocity O VI is detected). In the same figure, the distribution of high velocity H I 21 cm emission is also shown at the finer spatial resolution afforded by the H I data (see Wakker et al. 2003). Several large structures or groups of H I high velocity clouds are visible in Figure 2, including: 1) the Magellanic Stream, which passes through the south Galactic pole and extends up to $b \sim -30°$, with positive velocities for $l \gtrsim 180°$ and negative velocities for $l \lesssim 180°$; 2) high velocity cloud Complex C, which covers a large portion of the northern Galactic sky between $l = 30°$ and $l = 150°$ and has velocities of roughly –100 to –170 km s^{-1}; 3) the extreme positive velocity clouds in the northern sky (EPN), which are located in the general region $180^\circ \lesssim l \lesssim 330^\circ, b \approx 30^\circ$; and 4) the Galactic center negative velocity (GCN) clouds located near $l \sim 45^\circ, -60^\circ \lesssim b \lesssim -30^\circ$.

Several key points about the relationship of the high velocity O VI and H I in Figure 2 are worth highlighting:

1) There is excellent velocity correspondence between the H I and O VI in Complex C, with sight lines passing near to Complex C, but not through the high velocity H I, showing no O VI absorption.

2) Toward Complex A, which is near Complex C, there is a close pair of sight lines exhibiting an O VI detection (Mrk 106: $l = 161.^\circ1, b = 42.^\circ9^\circ$) and a non-detection (Mrk 116: $l = 160.^\circ5, b = 44.^\circ8$). At the 4–10 kpc distance of Complex A (Wakker 2001), the sight lines are separated by $\sim 140 - 350$ pc.

3) The high velocity component toward NGC 3310 ($l = 156.^\circ6, b = +54.^\circ1$) has a velocity similar to that of Complexes A and C, even though there is no H I 21 cm emission detected along the sight line at these velocities.

4) The H 1821+643 sight line ($l = 94.^\circ0, b = 27.^\circ4$) contains O VI absorption at the velocities of the Outer Arm as well as at more negative velocities. The progression of velocities between the Outer Arm and Complex C is relatively smooth.

5) In the $l < 180^\circ, b < 0^\circ$ quadrant of the sky there are often two negative velocity components, with the highest negative velocity components occurring in the Magellanic Stream or the extension of the Magellanic Stream at velocities typical of the Stream. Components with $\bar{v} \sim -120$ km s^{-1} concentrate to longitudes less than those of the Stream, whereas components at $l \sim 120^\circ - 140^\circ$ have velocities typical of the Stream.

6) There is good velocity correspondence between the H I and O VI velocities in the positive velocity portion of the Magellanic Stream at $l > 180^\circ$. The O VI features off the main axis of the Stream at these longitudes have velocities similar to those of the H I near the Stream.

7) In the $l > 180^\circ, b > 0^\circ$ quadrant of the sky there are positive velocity O VI and H I features, sometimes at similar velocities. In some cases, the O VI features have substantially higher velocities than the H I. The +277 km s^{-1} feature toward ESO 265-G23 ($l = 285.^\circ9, b = +16.^\circ6$) has a velocity and location close to that of H I in the leading arm of the Magellanic Stream identified by Putman et al. (1998); H I at similar velocities is seen $\sim 1^\circ$ away.

8) At $l \sim 180^\circ, b > 0^\circ$ there is high velocity O VI with $\bar{v} \sim +150$ km s^{-1}. Some of these features are broad absorption wings extending from the lower velocity absorption produced by the Galactic thick disk/halo.

9) There may be high velocity H I near the +143 km s^{-1} O VI feature toward PKS 0405-12 ($l = 204.^\circ9, b = -41.^\circ8$).

10) High velocity O VI features toward Mrk 478 ($l = 59.^\circ2, b = +65.^\circ0, \bar{v} \approx$

+385 km s^{-1}), NGC 4670 ($l = 212.°7, b = +88.°6, \bar{v} \approx +363$ km s^{-1}), and Ton S180 ($l = 139.°0, b = -85.°1, \bar{v} \approx +251$ km s^{-1}) stand out as having particularly unusual velocities compared to those of other O VI features in similar regions of the sky. These features may be located outside the Local Group (i.e., in the IGM).

11) Sight lines that contain both negative and positive high velocity features include Mrk 509 ($l = 36.°0, b = -29.°9$), Ton S180 ($l = 139.°0, b = -85.°1$), and several Complex C sight lines (PG 1259+593, PG 1351+640, Mrk 817, Mrk 876: $l \sim 85° - 120°, b \sim 40° - 60°$).

2.3 Column Densities

The high velocity O VI features have logarithmic column densities (cm^{-2}) of 13.06 to 14.59, with an average of $\langle \log N \rangle = 13.95 \pm 0.34$ and a median of 13.97 (see Figure 3, left panel). The average high velocity O VI column density is a factor of 2.7 times lower than the typical low velocity O VI column density found for the same sight lines through the thick disk/halo of the Galaxy (see Savage et al. 2003).

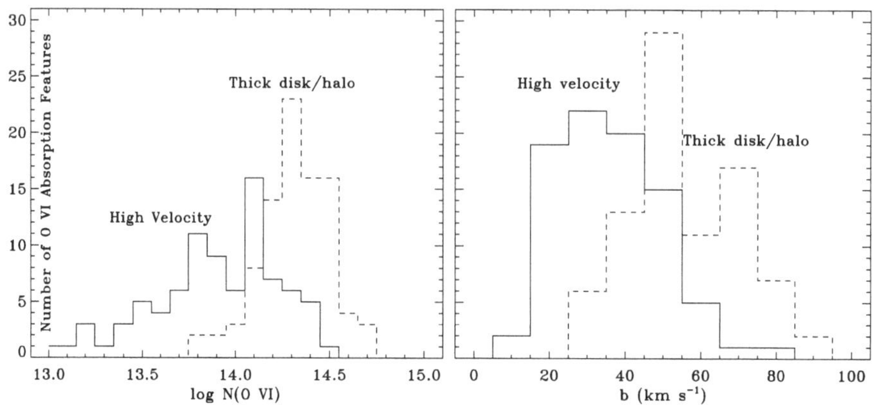

Figure 3. Histograms of the high velocity O VI column densities and line widths (solid lines). The bin sizes are 0.10 dex and 10 km s^{-1}, respectively. For comparison, the distributions for the O VI absorption arising in the thick disk and halo of the Galaxy are also shown (dashed lines) (from Sembach et al. 2003).

2.4 Line Widths

The line widths of the high velocity O VI features range from ~16 km s^{-1} to ~81 km s^{-1}, with an average of $\langle b \rangle = 40 \pm 14$ km s^{-1} (see Figure 3, right panel). The lowest values of b are close to the thermal width of 17.1 km s^{-1} expected for O VI at its peak ionization fraction temperature of $T = 2.8 \times 10^5$ K in collisional ionization equilibrium (Sutherland & Dopita 1993). The higher values of b require additional non-thermal broadening mechanisms or gas temperatures significantly larger than 2.8×10^5 K.

3. Origin of the High Velocity O VI

One possible explanation for some of the high velocity O VI is that transition temperature gas arises at the boundaries between cool/warm clouds of gas and a very hot ($T > 10^6$ K) Galactic corona or Local Group medium. Sources of the high velocity material might include infalling or tidally disturbed galaxies. A hot, highly extended ($R > 70$ kpc) corona or Local Group medium might be left over from the formation of the Milky Way or Local Group, or may be the result of continuous accretion of smaller galaxies over time. N-body simulations of the tidal evolution and structure of the Magellanic Stream favor a low-density medium ($n < 10^{-4}$ cm^{-3}) for imparting weak drag forces to deflect some of the Stream gas and providing a possible explanation for the absence of stars in the Stream (Gardiner 1999). Moore & Davis (1994) also postulated a hot, low-density corona to provide ram pressure stripping of some of the Magellanic Cloud gas. Hydrodynamical simulations of clouds moving through a hot, low-density medium show that weak bow shocks develop on the leading edges of the clouds as the gas is compressed and heated (Quilis & Moore 2001). Even if the clouds are not moving at supersonic speeds relative to the ambient medium, some viscous or turbulent stripping of the cooler gas likely occurs.

An alternative explanation for the O VI observed at high velocities may be that the clouds and any associated H I fragments are simply condensations within large gas structures falling onto the Galaxy. Cosmological structure formation models predict large numbers of cooling fragments embedded in dark matter, and some of these structures should be observable in O VI absorption as the gas passes through the $T = 10^5 - 10^6$ K temperature regime. The simulations suggest that $\sim$ 30% of the hot gas should be detectable in O VI absorption, while the remaining $\sim$ 70% may be visible in O VII and higher ionization stages (Davé et al. 2001).

The tenuous hot Galactic corona or Local Group gas may be revealed through X-ray absorption-line observations of O VII. The column density of O VII in the hot gas is given by $N(\mathrm{O\,VII}) = (\mathrm{O/H})_\odot\, Z\, f_{\mathrm{O\,VII}}\, nL$, where Z is the metallicity of the gas, f is the ionization fraction, and L is the path length, and $(\mathrm{O/H})_\odot = 5.45 \times 10^{-4}$ (Holweger 2001). At $T \sim 10^6$ K, $f_{\mathrm{O\,VII}} \approx 1$ (Sutherland & Dopita 1993). For $n = 10^{-4}$ cm^{-3}, $N(\mathrm{O\,VII}) \sim 2 \times 10^{16}\, Z\, (L/100$ kpc) (cm^{-2}). Preliminary results (Nicastro et al. 2002; Fang et al. 2003; Rasmussen et al. 2003) demonstrate that O VII absorption is detectable near zero velocity at a level consistent with the presence of a large, nearby reservoir of hot gas.

4. Are the O VI HVCs Extragalactic Clouds?

Some of the high velocity O VI clouds may be extragalactic clouds, based on what we currently know about their ionization properties. However, claims that essentially *all* of the O VI HVCs are extragalactic entities associated with an extended Local Group filament based on kinematical arguments alone appear to be untenable. Such arguments fail to consider the selection biases inherent in the O VI sample, the presence of neutral (H I) and lower ionization (Si IV, C IV) gas associated with some of the O VI HVCs, and the known "nearby" locations for at least two of the primary high velocity complexes in the sample — the Magellanic Stream is circumgalactic tidal debris, and Complex C is probably interacting with the Galactic corona. Furthermore, the O VII X-ray absorption measures used to support an extragalactic location have not yet been convincingly tied to either the O VI HVCs or to a Local Group location. The O VII absorption may well have a significant Galactic component in some directions (see Fang et al. 2003). The Local Group filament interpretation (Nicastro et al. 2003) may be suitable for some of the observed high velocity O VI features, but it clearly fails in other particular cases (e.g., the Magellanic Stream) or for the whole ensemble of high velocity O VI features in our sample. For example, the "Local Supercluster Filament" model (Kravtsov et al. 2002) predicts average O VI velocity centroids higher than those observed ($\langle \bar{v} \rangle \sim 1000$ km s^{-1} vs. $\langle \bar{v} \rangle < 400$ km s^{-1}) and average line widths higher than those observed (FWHM $\sim 100 - 400$ km s^{-1} vs. FWHM $\sim 30 - 120$ km s^{-1}). Additional absorption and emission-line observations of other ions at ultraviolet wavelengths would provide valuable information about the physical conditions, ionization, and locations of the O VI clouds.

References

Davé, R., Cen, R., Ostriker, J.P., et al. 2001, ApJ, 552, 473
Fang, T., Sembach, K.R., & Canizares, C.R. 2003, ApJ, 586, L49
Gardiner, L.T., 1999, in The Stromlo Workshop on High Velocity Clouds, ASP Conf. 166, eds. B.K. Gibson & M.E. Putman, (San Francisco: ASP), 292
Holweger, H. 2001, in Solar and Galactic Composition, AIP Conference Proceeding 598, ed. R.F. Wimmer-Schweingruber, (New York: American Institute of Physics), 23
Kravtsov, A., Klypin, A., & Hoffman, Y. 2002, ApJ, 571, 563
Lockman, F.J., Murphy, E.M., Petty-Powell, S., & Urick, V. 2002, ApJS, 140, 331
Moore, B., & Davis, M. 1994, MNRAS, 270, 209
Nicastro, F., Zezas, A., Drake, J., et al. 2002, ApJ, 573, 157
Nicastro, F., Zezas, A., Martin, E., et al. 2003, Nature, 421, 719
Quilis, V., & Moore, B. 2001, ApJ, 555, L95

Rasmussen, A., Kahn, S.M., & Paerels, F. 2003, astro-ph/0301183
Savage, B.D., Sembach, K.R., Wakker, B.P., et al. 2003, ApJS, May issue
Sembach, K.R., Wakker, B.P., Savage, B.D., et al. 2003, ApJS, May issue
Sutherland, R.S., & Dopita, M.A. 1993, ApJS, 88, 253
Wakker, B.P. 2001, ApJS, 136, 463
Wakker, B.P., Savage, B.D., Sembach, K.R., et al. 2003, ApJS, May issue

IONIZATION OF HIGH VELOCITY CLOUDS IN THE GALACTIC HALO

Jonathan D. Slavin
Harvard-Smithsonian Center for Astrophysics
jslavin@cfa.harvard.edu

Abstract Recent observations of Hα emission from high velocity clouds (HVCs) have raised the question of the ionization source for these clouds, which are believed to be fairly distant from ionizing radiation sources in the galactic disk. Obs ervations with FUSE of O VI absorption in HVCs also present us with questions regarding their source of ionization and the relation to the hot gas in the halo. We discuss sources for the ionization of the Hα emitting gas and the O VI containing gas. In particular we examine if interactions between warm ionized gas in the HVCs and the hot gas of the surrounding galactic halo could explain both the highly ionized gas and the ionization of cooler gas in the clouds.

1. Introduction

New *FUSE* and *WHAM* data show that gas with a wide range of ionization exists in high velocity clouds. In some cases the O VI absorption and Hα emission apparently come from the same cloud. These observations raise a number of interrelated questions.

- Assuming the Hα comes from warm, photoionized gas, what is the source of ionizing flux?
- Is the O VI gas photoionized or collisionally ionized?
- Is the O VI gas related to the Hα emitting gas and H I gas?
- What is the relationship of these HVCs to the hot Galactic halo gas?

The answers to these questions have the potential to tell us much about the Galactic halo and the intergalactic medium.

2. Warm Ionized Gas in HVCs

One surprise revealed by very sensitive Hα observations carried out with the Wisconsin Hα Mapper is the brightness of many HVCs. Hα emission has

R. Lieu and J. Mittaz (eds.), Soft X-Ray Emission from Clusters of Galaxies and Related Phenomena, 93–100.

been detected from clouds in complexes M, A and C (Tufte et al., 1998) and toward 4 of 5 compact HVCs observed (Tufte et al., 2002).

In nearly every direction looked at where there is significant high-velocity H I 21 cm emission, associated Hα emission has been found. Hα intensities for the HVCs are ~ 0.1 R ($= 10^5/4\pi$ photons cm^{-2} s^{-1} sr^{-1}) – if the emission is from optically thick clouds at $T \approx 10^4$ K that are photoionized, this implies a Lyman continuum flux of $\Phi_{\rm LyC} = 2.1 \times 10^5$ photons cm^{-2} s^{-1}. These Hα intensities are too high to be explained by photoionization from the metagalactic ionizing flux.

The warm ionized HVCs require a substantial Lyman continuum flux implying they are near a galactic source of ionizing radiation or have a diffuse source local to the halo.

3. Highly Ionized Gas in HVCs

FUSE has revealed that the HVCs contain highly ionized gas in addition to H I and the ionized hydrogen (Sembach et al., 2002). High velocity O VI ha s been seen in absorption in the majority of lines of sight observed towards extragalactic objects.

85 O VI features were observed including many associated with H I HVCs. Most of the O VI cannot be produced by photoionization because the required ionization parameter is too low (i.e. the density is too low). The size of regions implied by the required densities and column densities is too large for the regions to be coherent in velocity. The O VI is also very unlikely to be in quiescent hot gas since such gas either has little O VI (if $T \gtrsim 10^6$ K) or has a very high cooling rate (if $T \sim 10^{5.5}$ K). The existence of O VI HVCs associated with Hα and H I high-velocity gas points to a multi-phase structure in HVCs.

4. Sources for the Ionization of HVCs

Both the H^+ and the O^{+5} require substantial energy sources for their creation and maintenance in the Galactic halo. Being far from the Galactic disk requires either that the energy is transported from the disk or th at some sort of *in situ* source be present. Possible sources for photoionization include Lyman continuum flux from early type stars escaping the disk, diffuse emission from cooling hot gas in the disk and halo, the metagalactic ionizing flux, shocks between HVCs and halo gas, and EUV/soft X-ray emission at the boundaries between warm/cold clouds and hot halo gas. The latter is a potentially important *in situ* source that we discuss in detail below.

5. Sources for Hot Gas in the Halo

The O VI associated with high velocity H I almost certainly has as its energy source hot gas in the halo. Possible sources for the hot gas are superbubbles and supernovae that break out of the H I disk, accretion shocks during galaxy formation (part of the "warm hot intergalactic medium"), infall kinetic energy (again leading to shocks but in the current epoch), and a Galactic wind (most probably originating at the center of the Galaxy). The hot halo loses energy via radiative cooling and adiabatic expan sion.

6. Escape of Stellar Radiation

Several authors have estimated the fraction of O star radiation that leaks from the Galactic disk (e.g., Miller and Cox, 1993, Dove et al., 2000, Bland-Hawthorn and Maloney, 1999). These calculations depend on many factors including the scale height and morphology of the H I disk, the magnetic and cosmic ray pressure in the halo, superbubble evolution, and star formation history in OB associations. Consequently estimates for the escape fraction cover a broad range, $\sim 1 - 10\%$. The added unc ertainty of the distance to HVC gas adds to the uncertainty in whether escaping stellar radiation can account for its ionization.

These uncertainties and the association of Hα HVCs with O VI HVCs prompts us to look at the possibility that radiation generated in the interaction of hot halo gas and cooler gas could be a significant source of ionizing flux.

7. Ionizing Flux from Evaporative Cloud Boundaries

Evaporation of cool clouds via thermal conduction from surrounding hot gas has long be en proposed to occur in the Galactic disk and halo (e.g., Cowie and McKee, 1977). As the gas in such interfaces is heated and accelerated, it also emits prodigiously in the EUV. If this occurs around HVCs, it does not need to have a high luminosity in ionizing photons to provide significant ionization since it is generated in a thin layer that surrounds the warm gas. Because of this geometry a large fraction ($\sim$half) of the radiation is captured by the cloud.

For steady flow evaporation of a spherical cloud, th e energy equation is

$$\frac{1}{4\pi r^2}\frac{d}{dr}\left[\frac{5}{2}\dot{M}c^2 g + 4\pi r^2 q\right] = -L(T)n_e n_H, \qquad (1)$$

where $\dot{M}$ is the mass loss rate, c is the sound speed, $g = 1 + \mathcal{M}^2/5$ ($\mathcal{M}$ is the mach number), q is the (radial) heat flux and $L(T)$ is the radiative cooling coefficient. Inward heat flux is balanced by outward enthalpy flux reduced by cooling losses.

An important parameter in studies of evaporating clouds is the saturation parameter (see, e.g., Cowie and McKee, 1977),

$$\sigma_0 \equiv \frac{2\kappa_f T_f}{25\phi\rho_f c_f^3 R_{cl}} \propto \frac{T_f^3}{P R_{cl}}, \tag{2}$$

where κ_f is the classical (Spitzer) thermal conductivity at the asymptotic temperature, T_f, R_{cl} is the cloud radius, ϕ is the saturation fudge factor, c_f is the sound speed at T_f and P is the (thermal) pressure.

For highly saturated conduction (as occurs for high T_f and low P as expected in the halo) we find, using detailed numerical calculations of cloud evaporation, that the ionizing flux goes as

$$\Phi_{\rm LyC} \propto P^{1.1}, \tag{3}$$

with essentially no dependence on temperature of the hot gas (as long as σ_0 is high).

8. O VI from Evaporative Cloud Boundaries

In the boundary of an evaporating cloud O VI is generated as O is ionized from a low ionization state to beyond O^{+5} (assuming $T_f \gtrsim 10^{5.9}$ K). We expect $N(\rm O\,VI)$ $proptoM$ but $\dot{M}$ depends on T_f, R_{cl}, and σ_0. For highly saturated evaporation, we find a pressure dependence

$$N(\rm O\,VI) \propto P^{4/9} \tag{4}$$

Figure 1 shows the dependence of $\Phi_{\rm LyC}/N(\rm O\,VI)$ on σ_0 for a variety of assumptions for pressure, temperature and cloud radius. It can be seen that varying R_{cl} or varying T_h gives the same results dictated only by the value of σ_0 and P. For the specific case of a 3 pc radius cloud with $P = 5000$ cm^{-3} K, $T_f = 10^6$ K, and $Z = 1.0$, we find

$$N(\rm O\,VI) = 1.3 \times 10^{13}\ \rm cm^{-2} \tag{5}$$
$$\Phi_{\rm LyC} = 2.0 \times 10^4 \rm photons\ cm^{-2}\ s^{-1}. \tag{6}$$

Because the predicted values for $N(\rm O\,VI)$ and I_α are well below the observed values, each HVC would need to consist of ma ny ($\sim$ 10) evaporating clouds along a given line of sight which, given the angular size of the HVCs, is plausible. Calculations for lower pressures are difficult numerically because the degree of saturation becomes very high. The scaling of $N(\rm O\,VI)$ and I_α with P for runs of high saturation is very linear, however, giving us confidence in extrapolating to lower P values.

We have carried out limited testing of the metallicity dependence and have found that $\Phi_{\rm LyC}$ has a weak dependence on Z, $\sim Z^{0.06}$, while $N(\rm O\,VI) \propto Z$.

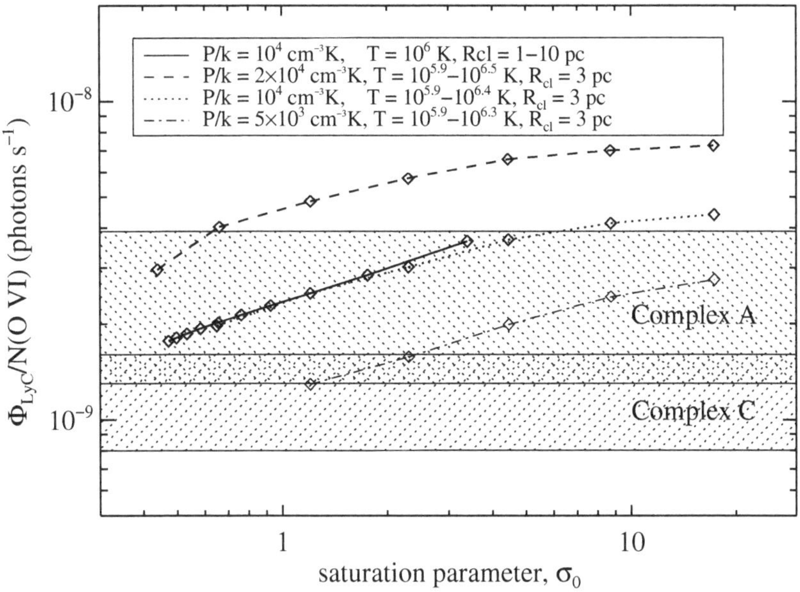

Figure 1. Dependence of the ratio, $\Phi_{\rm LyC}/N(\rm O\,VI)$, on the saturation parameter, σ_0, for evapo rating clouds. Shown are curves for a range of assumptions for the pressure, temperature and cloud radius. Note that the curves can be characterized by just two parameters, P and σ_0. The regions labeled Complex A and Complex C show the 1-σ uncertainty in the ratio as derived from Hα and O VI observations.

So $\Phi_{\rm LyC}/N(\rm\ O\,VI)$ goes roughly as $\sim Z^{0.94}$. The weak dependence of $\Phi_{\rm LyC}$ on Z is in part due to the fact that a large fraction of the ionizing flux comes from the He II 304Å line and in part due to the complex interplay of non-equilibrium ionization effects and cooling in cloud boundaries. In general the gas in evaporative outflows is significantly under-ionized relative to collisional ionizati on equilibrium (CIE). Under-ionized gas tends to have a substantially higher emissivity than CIE gas. Also, the evaporation rate is decreased by radiative cooling in the flow. Thus there is something of a self-regulation process wherein for lower Z clouds, the radiative cooling would be expected to be lower, yet the reduced cooling rate allows for a higher mass loss rate which in turn leads to a flow that is farther from CIE and thus has enhanced radiative cooling. The net result is a weak dependence on Z of the cloud boundary temperature profile, mass loss rate and total ionizing flux.

9. An Alternative Interface Model: Turbulent Mixing Layers

Another, and possibly more likely form for the interaction of hot halo gas and cool HVCs is in turbulent mixing layers (TML). The TML model (Begelman and Fabian, 1990; Slavin et al., 1993) proposes that shear flows between hot gas and cool clouds results in Kelvin-Helmholtz instabilities. Turbulent mixing rapidly occurs in the shear layer leading to gas at an intermediate temperatu re, $\bar{T}$, between the hot gas temperature, T_h, and the cloud temperature, T_c. After mixing the gas rapidly cools and radiates. O VI is generated in the cooling gas as is EUV radiation. Given the high relative speed of HVCs and gas coupled to Galactic rotation, substantial shear is likely.

We have further explored models of the type proposed by Slavin et al., 1993 with an eye to their application to the ionization of HVCs. In these TML models the layer is assumed to have reached a steady state i n which cooling in the layer balances enthalpy flux into the layer. As a result, $\Phi_{\rm LyC} \propto P$. The flux depends on $\bar{T}$, but does not depend sensitively on T_h. O VI is independent of P, (due to the steady state assumption) but depends strongly on both $\bar{T}$ and T_h. Because the metal lines, particularly O VI, are important coolants, $N(\rm O\,VI)$ has a weak dependence on Z. TMLs produce relatively small amounts of O VI and $\Phi_{\rm LyC}$ per layer, but we exp ect the clouds to contain many such layers – each HVC would be a cluster of cloudlets that are being turbulently disrupted.

10. Comparison with the Data

To date we have analyzed two cases in which there is Hα emission observed and O VI absorption that are clearly associated in an HVC, though there are potentially many more. For Complex C we have $I_\alpha = 0.13 \pm 0.03$ R which implies $\Phi_{LyC} = 2.7 \times 10^5$ cm^{-2} s^{-1} and $N(\rm O\,VI) = (2.2 \pm 0.6) \times 10^{14}$ cm^{-2} which yields $\Phi_{LyC}/N(\rm O\,VI) = (1.2 \pm 0.4) \times 10^{-9}$ photons s^{-1}.

For Complex A there is a weak O VI detection, $N(\rm O\,VI) = (6 \pm 3) \times 10^{13}$ cm^{-2}, and we have $I_\alpha = 0.08 \pm 0.01$ R or $\Phi_{LyC} = 1.6 \times 10^5$ cm^{-2} s^{-1} giving $\Phi_{LyC}/N(\rm O\,VI) = (2.6 \pm 1.3) \times 10^{-9}$ photons s^{-1}.

Both directions are consistent with $\Phi_{LyC}/N(\rm O\,VI) \sim (1{-}2) \times 10^{-9}$photons s^{-1}. Both the cloud evaporation model and the TML model can attain this value for moderately low pressures, $P/k \sim 1000$ cm^{-3} K.

11. Interface Radiation as a Source of Ionizing Radiation

While the detailed predictions for Φ_{LyC} and $N(\rm O\,VI)$ depend on the paricular model and its parameters, there are some general conclusions we can draw from the existence of O VI HVCs.

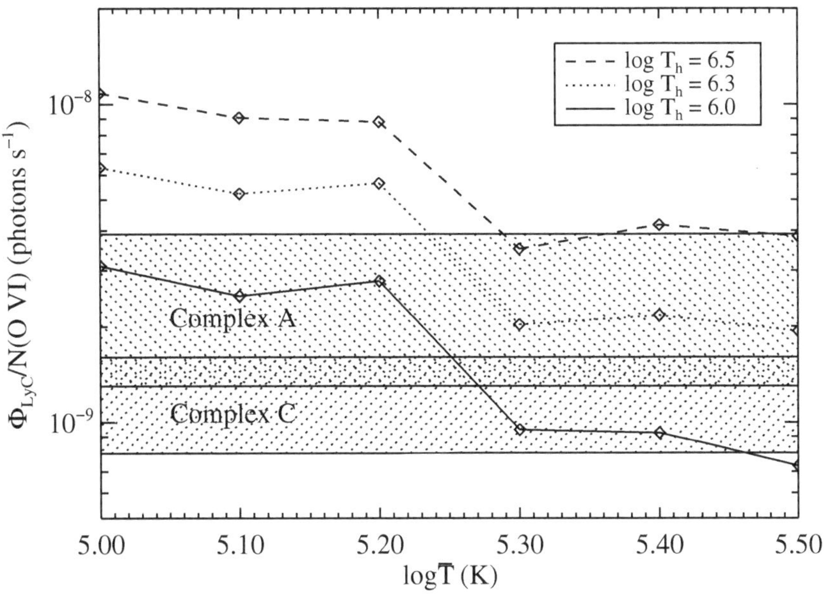

Figure 2. Dependence of the ratio, $\Phi_{\rm LyC}/N(\rm O\,VI)$, on the mixing temperature, $\bar{T}$, for turbulent mixing layers. The various curves are for different values of the hot gas temperature. For all cases the $T_c = 10^4$ K and $P/k = 1000$ cm^{-3} K.

- Radiation from O VI containing gas produces significant amounts of H ionizing radiation.

- The existence of O VI in HVCs indicates either hot gas cooling or cool gas being heated and thus we should expect warm photoionized gas where we see O VI.

- Because neutral HVCs can carry the ir ionizing radiation source with them, the linkage between the radiation field of the Galactic disk and the ionizing flux seen by the clouds is broken. The bad news is that we cannot tell the distance to the HVCs based on their Hα emission. The good news is that we have a potential explanation for the ionization of all HVCs.

- The flux from interfaces is unfortunately model dependent and depends on the pressure, temperature and abundances in the HVCs and hot gas.

12. Conclusions

- The HVCs that have associated H I and Hα emission appear to be in the Galactic halo and part of a multi-phase extended hot halo.
- Interfaces between hot halo gas and the cooler H I/H II clouds can easily produce the required Lyman continuum fluxes needed to provide the ionization of H; the ratio, $\Phi_{LyC}/N(\mathrm{O\,VI}) \sim (1-2)\times 10^{-9}$ photons s^{-1}, can be matched both in evaporating cloud models and in turbulent mixing layer models.
- Observations of Hα on more lines of sight with observed O VI and observations of more ions on those lines of sight are needed to constrain the nature of the multi-phase HVC clouds and their relationship to hot halo gas and the IGM.

References

Begelman, M. C. and Fabian, A. C. (1990). Turbulent mixing layers in the interstellar and intracluster medium. *MNRAS*, 244:26P.

Bland-Hawthorn, J. and Maloney, P. R. (1999). The Escape of Ionizing Photons from the Galaxy. *ApJ*, 510:L33–L36.

Cowie, L. L. and McKee, C. F. (1977). The evaporation of spherical clouds in a hot gas. i - classical and saturated mass loss rates. *ApJ*, 211:135.

Dove, J. B., Shull, J. M., and Ferrara, A. (2000). The Escape of Ionizing Photons from OB Associations in Disk Galaxies: Radiation Transfer through Superbubbles. *ApJ*, 531:846–860.

Miller, W. W. I. and Cox, D. P. (1993). The Diffuse Ionized Interstellar Medium: Structures Resulting from Ionization by O Stars. *ApJ*, 417:579.

Sembach, K. R., Wakker, B. P., Savage, B. D., Richter, P., Meade, M., Shull, J. M., Jenkins, E. B., Moos, H. W., and Sonneborn, G. (2002). Highly-Ionized High-Velocity Gas in the Vicinity of the Galaxy. *ApJS*, in press (astro-ph/0207562).

Slavin, J. D., Shull, J. M., and Begelman, M. C. (1993). Turbulent mixing layers in the interstellar medium of galaxies. *ApJ*, 407:83.

Tufte, S. L., Reynolds, R. J., and Haffner, L. M. (1998). WHAM Observations of H alpha Emission from High-Velocity Clouds in the M, A, and C Complexes. *ApJ*, 504:773.

Tufte, S. L., Wilson, J. D., Madsen, G. J., Haffner, L. M., and Reynolds, R. J. (2002). WHAM Observations of Hα from High-Velocity Clouds: Are They Galactic or Extragalactic? *ApJ*, 572:L153–L156.

III

SOFT X-RAY DATA ANALYSIS ISSUES

THE ISM FROM THE SOFT X-RAY BACKGROUND PERSPECTIVE

S. L. Snowden
NASA/GSFC and USRA
snowden@riva.gsfc.nasa.gov

Abstract In the past few years progress in understanding the local and Galactic ISM in terms of the diffuse X-ray background has been as much about what hasn't been seen as it has been about detections. High resolution spectra of the local SXRB have been observed, but are inconsistent with current thermal emission models. An excess over the extrapolation of the high-energy (most clearly visible at $E >$ 1.5 keV) extragalactic power law down to $\frac{3}{4}$ keV has been observed but only at the level consistent with cosmological models, implying the absence of at least a bright hot Galactic halo. A very recent *FUSE* result indicates that O VI emission from the Local Hot Bubble is insignificant, if it exists at all, a result which is also inconsistent with current thermal emission models. A (very) short review of the Galactic SXRB and ISM is presented.

1. Introduction

What is the soft X-ray diffuse background (SXRB), at least in relevance to the Galactic interstellar medium? Phenomenologically, the SXRB is whatever is left over after all other distinct identified emission sources have been removed. Bright point sources (or effective point sources) such as AGN, binary systems, black holes, and pulsars are first excised from the data. Next, bright extended structures such as clusters of galaxies and young supernova remnants which stand out from the background as distinct individual objects are removed, objects such as the Virgo Cluster, Cygnus Loop supernova remnant (SNR), and the Vela SNR. Thus far the winnowing process is simple with the only complication being determining the angular extent and brightness at which an object becomes a "bright extended structure," and so is removed. The next step is to account for unresolved objects which are contributing to the residual background, primarily stars and AGN, which is where the situation does get complicated.

The residual background component made up of cosmological objects (primarily AGN) is most clearly observed at energies above 1.5 keV where Galac-

R. Lieu and J. Mittaz (eds.), Soft X-Ray Emission from Clusters of Galaxies and Related Phenomena, 103–110.

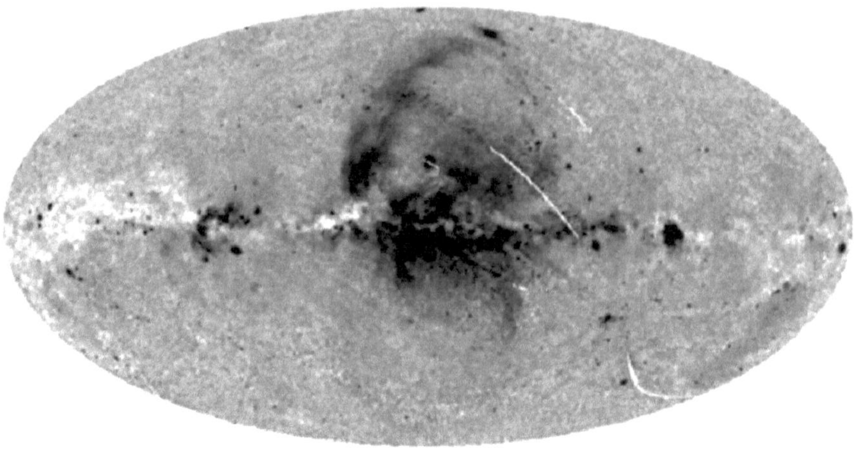

Figure 1. *ROSAT* All-Sky Survey (RASS) image of the $1-2$ keV band SXRB (Snowden et al. 1997), displayed with an Aitoff-Hammer zero-centered projection in Galactic coordinates. The data have been square-root scaled, longitude increases to the left, and darker shading indicates brighter emission. Note the absorption minimum along the Galactic plane, the Loop I and Galactic bulge emission in the direction of the Galactic center and rising to high latitudes. The North Polar Spur (NPS) is the limb-brightened edge of the Loop I emission running diagonally north and then west from near the plane at $l \sim 30°$ to $l, b \sim 300°, 70°$. Several bright but smaller objects are visible along the Galactic plane, e.g., the Cygnus Superbubble at $l \sim 90°$ and the Vela SNR at $l \sim 270°$. The Coma cluster is the small enhancement at the very top of the image while the Virgo cluster is the larger enhancement at the top of the NPS. The Large Magellanic Cloud is the small emission region at $l, b \sim 270°, -30°$.

tic emission is minimal. Except for absorption modulation by material of the Galactic disk and Loop I/Galactic bulge emission toward the Galactic center it appears isotropic (Figure 1). There are several groups which have worked rather successfully on resolving this background by the use of deep surveys (e.g., Hasinger et al. 1993; Mushotzky et al. 2000; Giacconi et al. 2001). This is convenient for those of us studying the Galactic SXRB as they provide highly accurate models for what must be subtracted to leave the truly interesting signal. Unresolved Galactic stars also provide a background component but fortunately at a relatively low level which primarily affects the $\frac{3}{4}$ keV band (Kuntz and Snowden 2001b).

The observed Galactic SXRB covers the energy range $\sim 0.1 - 1.5$ keV and originates primarily as diffuse emission from thermal processes with temperatures of $kT \sim 0.1 - 1$ keV. The lower limit of the energy band is set by technical reasons (typical detectors for observing the SXRB have very small responses approaching 0.1 keV) and by definition (the X-ray and EUV bands meet at about 100 eV). The upper limit is more nebulous being set by the emission mechanisms, however even the higher temperature extended Galactic emission is still less than $kT \sim 1$ keV.

2. A Short Guided Tour of the SXRB

Away from the Galactic center quadrant the SXRB in the $1 - 2$ keV band (Figure 1) is relatively isotropic illustrating the contaminating contribution of unresolved (and therefore unremoved) extragalactic sources. Emission from the Galactic bulge and Loop I (and most clearly the NPS) dominate the Galactic center region. The Cygnus Superbubble at $l \sim 90°$ and the Vela SNR at $l \sim 270°$ are clearly seen along the Galactic plane. The effect of Galactic absorption is seen most clearly in the plane between $l \sim 90°$ and $l \sim 180°$.

Similar to the $1 - 2$ keV band, the $\frac{3}{4}$ keV band is dominated by the unresolved extragalactic background except in the Galactic center quadrant and along the Galactic plane. However, there are a few Galactic objects which become more visible (e.g., the Eridanus Superbubble and Monoceros/Gemini SNR, the latter also known as the Monogem Ring) as the energy range of the band becomes more compatible with the emission temperatures. Galactic absorption features also become clearer as the disk becomes optically thick at higher latitudes (ISM absorption cross sections go roughly as $E^{-\frac{8}{3}}$).

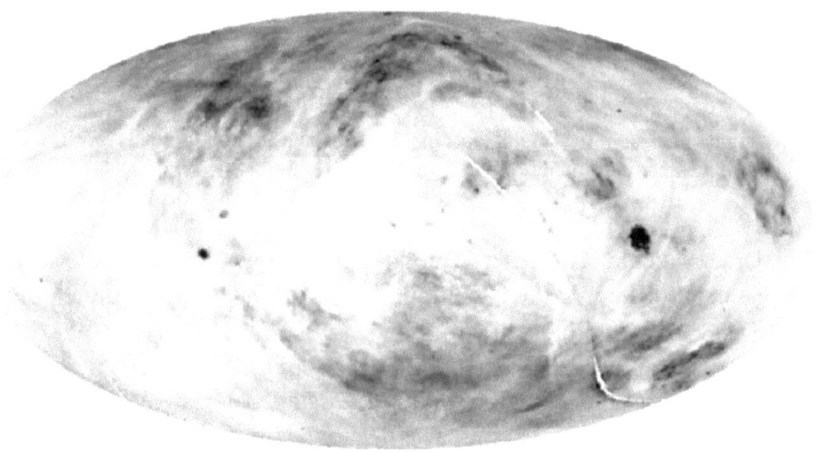

Figure 2. Same as Figure 1 except for the $\frac{1}{4}$ keV band and the data have been linearly scaled. Galactic absorption is optically thick to much higher latitudes ($|b| \sim 30°$) and obscures all but the nearest objects in the Galactic plane. However, a number of Galactic features have become readily apparent, e.g., the Monoceros/Gemini SNR at $l, b \sim 200°, 10°$, the Eridanus (Eridion) Superbubble at $l, b \sim 190°, -30°$, and the Cygnus Loop SNR at $l, b \sim 90°, -5°$. The more northerly part of the NPS is still visible but the southern end is obscured by absorption.

The structure of the $\frac{1}{4}$ keV background (Figure 2) is completely different from the higher energy bands due both to additional Galactic emission and the stronger effect of Galactic absorption. This is where the subject of this paper becomes critical for the goals of this conference. The Galactic $\frac{1}{4}$ keV background is much brighter than the extrapolation of the isotropic extragalactic background, can and does vary significantly on angular scales of a degree or less, and of perhaps more importance for the subject of this conferences can play merry havoc with observations of soft X-ray emission from clusters of Galaxies.

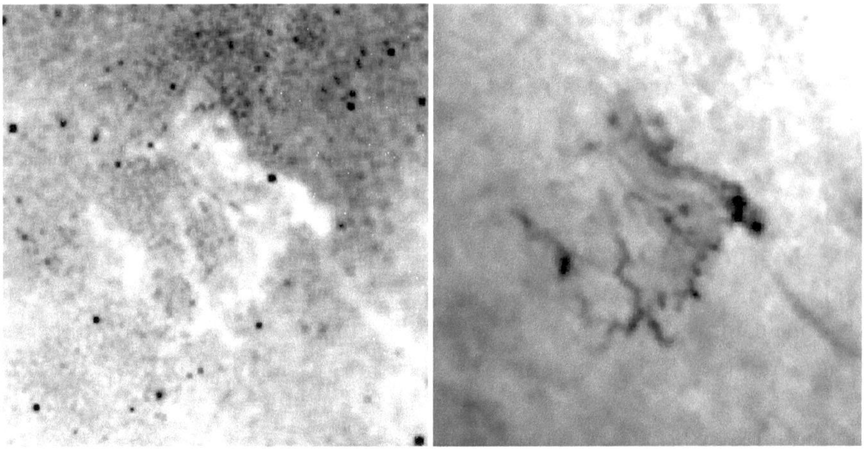

Figure 3. RASS $\frac{1}{4}$ keV (left) and *IRAS* 100 μm (right) images of the Draco Nebula. The fields are the same and are $12.8^\circ \times 12.8^\circ$ in extent. Galactic north is up and darker shading indicates higher intensity. The bright knots in the X-ray data are point sources of various types. Note the detailed and fine structure of the X-ray data.

The Draco Nebula provides an exceptional example of the possible structure of the Galactic diffuse X-ray background at $\frac{1}{4}$ keV, and the coupling between the ISM and that structure. Figure 3 shows both the $\frac{1}{4}$ keV *ROSAT* All-Sky Survey (RASS) and cleaned *IRAS* 100 μm (Schlegel, Finkbeiner, and Davis 1998) data for the Draco region where the detailed negative correlation between the two data sets is striking. The negative correlation also illustrates the primary tool for locating the emission regions in space by use of the "shadowing" technique. By looking at the detailed negative correlation, i.e., the variation of SXRB intensity with absorbing column density, it is possible to model the foreground and background (to the absorbing ISM) X-ray emission. With the knowledge of the distance to the absorbing material it is possible to locate the emission in three-dimensional space. For the case of the Draco Nebula, the distance and latitude places it at the upper edge of the Galactic disk implying the existence of strong $\frac{1}{4}$ keV emission in the Galactic halo, as well as significant emission in the nearest few hundred parsecs.

3. The Hot Phase of the Galactic ISM

Now for the questions: What is the hot phase of the Galactic ISM? Where is it located? How much of it is there? Where did it come from?

What is it? The hot ISM is the remnant of various energetic processes in the Galaxy such as supernovae and stellar winds. The X-ray emitting plasmas are diffusely distributed over volumes extending hundreds of parsecs, and in the case of the Galactic bulge several kiloparsecs. The plasmas have temperatures of $kT \sim 0.1 - 1$ keV and correspondingly low space densities ($n_e \sim 0.01$ cm^{-2}). Recent observations have confirmed that at least the $\frac{1}{4}$ keV emission is thermal in nature and dominated by various lines (Sanders et al. 2001). The observations also showed that the emission is inconsistent with that from a plasma in thermal equilibrium with solar abundances.

Where is the $\frac{1}{4}$ keV emission? Making use of the RASS and cleaned *IRAS* 100 μm data in a shadowing analysis it has been possible to produce a picture of at least the local (out to a few hundred parsecs in the plane, farther in the halo) $\frac{1}{4}$ keV emission (Snowden et al. 1998). From ISM optical absorption line studies the Sun is located in a cavity in the H I of the Galactic disk extending from less than ~ 50 pc from the Sun in the plane to well over 100 pc at high Galactic latitudes (e.g., Sfeir et al. 1999). From the X-ray data it is clear that this cavity contains an extensive distribution of thermal plasma at $kT \sim 0.1$ keV, which has been given the name the Local Hot Bubble (LHB). A few distinct emission regions such as the Monoceros/Gemini SNR, Eridanus Superbubble, Vela SNR, and Cygnus Loop SNR mentioned above can be seen near the plane out to several hundred parsecs, but that is about the limit as ISM absorption cuts off observations of emission regions at greater distances. At higher Galactic latitudes absorption optical depths for $\frac{1}{4}$ keV X-rays drop significantly to a minimum of $\tau \sim 0.5$ but with typical values from $\sim 1 - 1.5$. The X-ray emission in the halo appears rather patchy with intensities ranging from near zero to many times typical values for the LHB.

Where is the $\frac{3}{4}$ keV emission? The two largest and brightest $\frac{3}{4}$ keV Galactic emission regions as seen from the Earth unfortunately lie in the same direction, that of the Galactic center. The Loop I superbubble is powered by the Sco-Cen OB associations and has a radius of ~ 100 pc and is centered at a distance of ~ 150 pc in the direction of $l, b \sim 330°, 15°$. The Galactic bulge is most clearly seen in the southern hemisphere where it is not confused with the Loop I emission. It appears to have a radial extent of ~ 5 kpc, a scale height of ~ 2 kpc, a temperature of $kT \sim 0.4 - 0.5$ keV, and a luminosity of $\sim 2 \times 10^{39}$ ergs s^{-1} (Snowden et al. 1997). It is not clear whether there exists a large scale-height $\frac{3}{4}$ keV halo such as suggested by the work of Wang (1998). Kuntz and Snowden (2001a) and Kuntz, Snowden, and Mushotzky (2001) place restrictive limits on the amount of halo and truly extragalactic emission in this band. Any such extended halo emission is going to be confused with more cosmological emission, at least to the extent of the Local Group of galaxies. A recent high-resolution spectrum of the diffuse background places strong limits on the spectral distribution of the $\frac{3}{4}$ keV background based on the O VII and O VIII emission lines (McCammon et al. 2002). The oxygen results when combined with the deep survey AGN spectra show that 42% of the observed flux can be attributed to unresolved AGNs and a minimum of 36% to thermal emission at zero redshift (which would include the contributions of both unresolved stars and the cooler plasma contributing to the $\frac{1}{4}$ keV background). Subtracting the high-latitude contribution of unresolved stars, LHB, and cooler halo leaves an upper limit of $\sim$ 20% of the observed flux which (because of its low redshift) has either a Galactic halo or local group origin.

How much of it is there? The emission at $\frac{3}{4}$ keV from the Galactic bulge is well constrained, as is the hotter Galactic halo, at least by upper limits (see above). At $\frac{1}{4}$ keV the amount of emission from the LHB is fairly well constrained as is the nearest few hundred parsecs to kiloparsec of the Galactic halo. However, the amount of $kT \sim 0.1$ keV plasma in the disk of the Milky Way as a whole is essentially unknown as it is invisible to us. The fact that the $\frac{1}{4}$ keV halo appears so patchy suggests that it lies at a relatively low scale height and presumably just above the neutral gas of the disk. If this is the case then we know little more about the extent of the cooler halo as we can only sample $\sim$ 1% of the Galaxy.

Where did it come from? The bottom line is that the production of extensive plasma with temperatures in the $kT \sim 0.1 - 1$ keV range requires very energetic phenomena such as supernovae and the stellar winds of OB associations and star forming regions. This is clearly the case for many of the distinct objects mentioned above such at the Loop I superbubble powered by the Sco-Cen OB associations and the Eridanus Superbubble powered by the Orion OB associations. The Galactic bulge and the hotter part of the Galactic halo (if it exists) may be powered by star formation and perhaps infall from the IGM. The cooler part of the Galactic halo can be produced either by the breakout of emission regions in the disk (otherwise known as Galactic fountains) or in-situ by halo supernovae. The LHB appears to be an aging supernova remnant with no candidate for the originating object. The physical extent of the local cavity suggests that in was created through the effort of more than one supernova, and plasmas at $kT \sim 0.1$ keV are relatively long lived with cooling times of millions of years providing a time scale where that would not be unreasonable. Maíz-Apellániz (2001) suggests that a component of the Sco-Cen OB associations passed through the region of the local cavity several million years ago and supernovae at that time could have created or reheated the plasma in an existing cavity. The occurrence of a supernova within the last five million years and within 30 pc of the Sun is supported by the results of Knie et al. (1999) who analyzed ^{60}Fe (a supernova byproduct) in ocean-floor sediment.

4. Caveats and Other Considerations

The McKee and Ostriker model for the ISM: I raise this subject as there seemed to be some surprise at the conference that although the model had a good run it has been effectively ruled out. McKee and Ostriker (1977) postulated an ISM comprised of a highly clumped cooler component (H I clouds) embedded in a hotter matrix (X-ray emitting plasma) where the O VI observed by Copernicus could arise from the interface region between the two. This model solved a number of model/data inconsistencies at the time and proved rather popular over the years. However, such a clumped distribution of cold clouds has been searched for but has not been observed over all reasonable angular scales in 21-cm emission and absorption. In addition, the X-ray data do not show any support for the interconnecting matrix of hot plasma. (See Snowden 2001 for a somewhat more extended discussion of this subject).

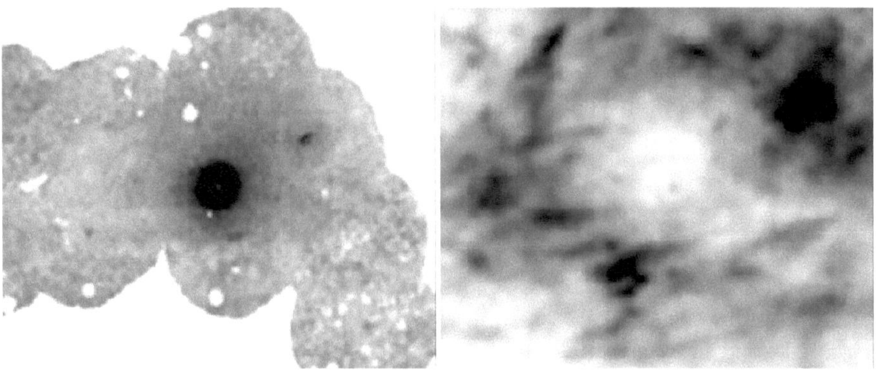

Figure 4. $ROSAT$ $\frac{1}{4}$ keV (left) and $IRAS$ 100 μm (right) images of the Virgo Cluster. The fields are the same and are $\sim 6° \times 5°$ in extent. Darker shading indicates higher intensity. The peak $IRAS$ 100 μm intensities correspond to several optical depths at $\frac{1}{4}$ keV.

The pathological nature of the Virgo Cluster region: This digression is included as a cautionary tale about the analysis of soft X-ray data and the effects of ISM absorption. Figure 4 shows the $\frac{1}{4}$ keV and $IRAS$ 100 μm images of the Virgo Cluster region, the most pathological case I know of where ISM absorption conspires to confuse the analysis of a completely unrelated object. The ring of $IRAS$ cirrus surrounds the brightest region of the cluster emission with $\sim 1-2$ optical depths of additional absorption at $\frac{1}{4}$ keV. Because the cirrus ring is roughly circular as well as roughly centered on the cluster it significantly modifies the radial profile of the $\frac{1}{4}$ keV emission. However, the additional absorption is relatively small at higher energies so the radial profile of, for instance, the $\frac{3}{4}$ keV to $\frac{1}{4}$ keV hardness ratio is also affected.

Where has all the O VI gone? Shelton (2003) presents results which throw some confusion on models for the LHB and cold cloud/hot plasma interfaces. Using *FUSE* observations there appears to be little or no O VI emission from the LHB where there should be $T \sim 300,000$ K gas in interface regions between the hotter plasma and the cooler material of the cavity walls. There should also be O VI emission from the edge of small clouds located within the cavity (the Sun itself is located in the "Local Fluff," a region of partially ionized gas at a few thousand degrees with an extent of a few parsecs).

Current spectral models don't fit the observed $\frac{1}{4}$ keV spectra: While there are a few relatively good spectra of the diffuse background at $\frac{1}{4}$ keV they are not well fit by current thermal equilibrium and normal abundance emission models, or non-equilibrium models for that matter (Sanders et al. 2001).

What is the zero level of the $\frac{1}{4}$ keV X-ray data? Recently the production of soft X-rays in the solar system by charge exchange between the solar wind and interstellar neutrals has become a hot topic. While so far this mechanism has been used to explain the flaring long-term enhancement background observed in the RASS (Cravens, Robertson, and Snowden 2001), the question arises whether at quiescence it still produces a non-negligible flux which could significantly impact our view of the LHB (e.g., Lallement 2003).

References

Cravens, T. E., Robertson, I. P., and Snowden, S. L. 2001, JGR, 106, 24883

Giacconi, R., et al. 2001, ApJ, 551, 624

Hasinger, G., et al. 1993, A&A, 275, 1

Knie, K., et al. 1999, PhRvL, 83, 18

Kuntz, K. D., and Snowden, S. L. 2001a, ApJ, 543, 195

Kuntz, K. D., and Snowden, S. L. 2001b, ApJ, 554, 684

Kuntz, K. D., Snowden, S. L., and Mushotzky, R. F. 2001, ApJL, 548, 119

Lallement, R. 2003, A&A, submitted

Maíz-Apellániz, J. 2001, ApJL, 560, L83

McCammon, D., et al. 2002, ApJ, 576, 188

McKee, C. F., and Ostriker, J. P. 1977, ApJ, 218, 148

Mushotzky, R. F., Cowie, L. L., Barger, A. J., and Arnaud, K. A. 2000, *Nature*, 404, 459

Sanders, W. T., et al. 2001, ApJ, 554, 694

Schlegel, D. J., Finkbeiner, D. P., and Davis, M. 1998, ApJ, 500, 525

Sfeir, D. M., Lallement, R., Crifo, F., and Welsh, B. Y. 1999, A&A, 346, 785

Shelton, R. 2003, ApJ, in press

Snowden, S. L. 2001, in *The Century of Space Science*, Klewer Academic Publishers, 581

Snowden, S. L., et al. 1997, ApJ, 485, 125

Snowden, S. L., et al. 1998, ApJ, 493, 715

Wang, Q. D. 1998, Lecture Notes in Physics, 506, 503

PEERING THROUGH THE MUCK: NOTES ON THE INFLUENCE OF THE GALACTIC INTERSTELLAR MEDIUM ON EXTRAGALACTIC OBSERVATIONS

Felix J. Lockman
*National Radio Astronomy Observatory**
Green Bank, WV USA
jlockman@nrao.edu

Abstract This paper considers some effects of foreground Galactic gas on radiation received from extragalactic objects, with an emphasis on the use of the 21cm line to determine the total N_{HI}. In general, the opacity of the 21cm line makes it impossible to derive an accurate value of N_{HI} by simply applying a formula to the observed emission, except in directions where there is very little interstellar matter. The 21cm line can be used to estimate the likelihood that there is significant H_2 in a particular direction, but carries little or no information on the amount of ionized gas, which can be a major source of foreground effects. Considerable discussion is devoted to the importance of small-scale angular structure in HI, with the conclusion that it will rarely contribute significantly to the total error compared to other factors (such as the effects of ionized gas) for extragalactic sight lines at high Galactic latitude. The direction of the Hubble/Chandra Deep Field North is used as an example of the complexities that might occur even in the absence of opacity or molecular gas.

1. Introduction

The Interstellar Medium (ISM) regulates the evolution of the Galaxy. It is the source of material for new stars and the repository of the products of stellar evolution. But it is a damned nuisance to astronomers seeking to peer beyond the local gas. In this article I treat the ISM as if it were simply an impediment to knowledge, and suggest ways that one might estimate its effects. This topic has taken on increasing importance in recent years as more and more experiments are requiring correction for the "Galactic foreground" (e.g., Hauser

*The National Radio Astronomy Observatory is operated by Associated Universities, Inc., under a cooperative agreement with the National Science Foundation.

R. Lieu and J. Mittaz (eds.), Soft X-Ray Emission from Clusters of Galaxies and Related Phenomena, 111–121.

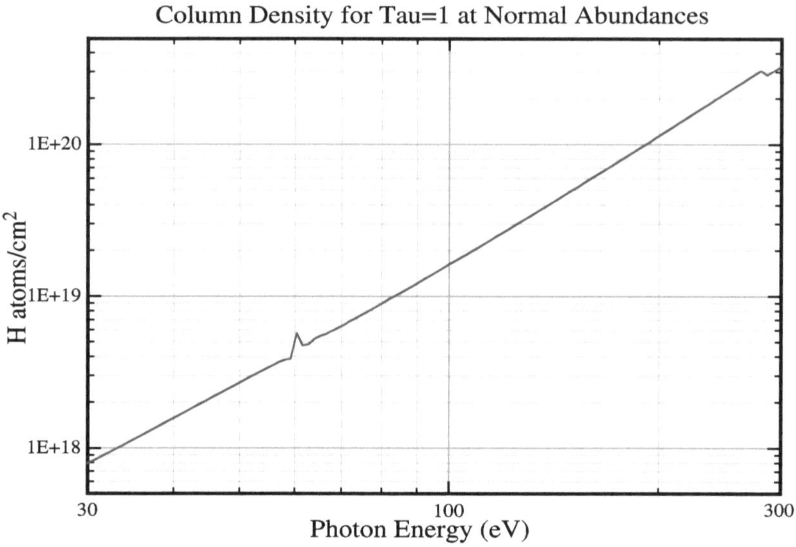

Figure 1. The equivalent N_{HI} needed to produce an opacity of unity for the Galactic ISM at normal abundances, as a function of photon energy (Balucinska-Church & McCammon 1992).

2001). Here the emphasis will be on the use of the 21cm line to determine a total N_{HI}, for the 21cm line is our most general tool, and N_{HI} is an important quantity which can be used to estimate N_{He}, E(B–V) and $S_{100\mu}$, as well as the likelihood that there is molecular hydrogen along the line of sight. Some of the points treated here are discussed in more detail in reviews by Kulkarni & Heiles (1987), Dickey & Lockman (1990; hereafter DL90), Dickey (2002), and Lockman (2002).

2. General Considerations

Figure 1 shows the amount of neutral interstellar gas, expressed as an equivalent HI column density, needed to produce unity opacity given normal abundances. Below the C-band edge at 0.25 KeV the opacity results almost entirely from photoelectric absorption by hydrogen and helium, which contribute about equally to τ (Balucinska-Church & McCammon, 1992).

Surveys of the sky in the 21cm line find $\langle N_{HI} sin|b| \rangle = 3 \times 10^{20}$ cm^{-2} where b is the Galactic latitude (DL90). Thus, *most* sight-lines through the Milky Way have $\tau \geq 1$ for $13.6 < E < 300$ eV, even without taking into account any contribution to the opacity from molecular hydrogen, H_2. Luckily, there are large areas of the sky over which N_{HI} is a factor of several below the average, but even so, the lowest N_{HI} in *any* known direction is 4.4×10^{19} cm^{-2} (Lockman, Jahoda & McCammon 1986; Jahoda, Lockman & McCammon 1990), so observations at $13.6 < E \leq 100$ eV must always involve substantial corrections for the Galactic ISM.

Except in directions where it is possible to make a direct measurement of N_{HI} in UV absorption lines, every attempt to determine the effect of the ISM on extragalactic observations should begin with the 21cm line. The Leiden-Dwingeloo (LD) 21cm survey covering $\delta > -30^\circ$ at $35'$ angular resolution (Hartmann & Burton 1997) supersedes all previous general surveys because of its angular and velocity resolution, and high quality of data. A southern extension will be completed soon. Some parts of the sky of special interest have been mapped at higher angular resolution (e.g., Elvis et al. 1994; Miville-Deschênes et al. 2002; Barnes & Nulsen 2003). The brighter emission near the Galactic plane is now being measured at $1'$ resolution by a consortium who employ three different synthesis arrays (Knee 2002; McClure-Griffiths 2002; Taylor et al. 2002).

3. Estimating N_{HI} from 21cm HI Data

Radio telescopes measure an HI brightness temperature, T_b, as a function of velocity, but in general, there is no single formula that can be applied to derive N_{HI} from the observed T_b. The solution to the equation of transfer for 21cm emission from a uniform medium is simple enough:

$$T_b(v) = T_s[1 - exp(-\tau(v))], \tag{1}$$

where $\tau(v) = 5.2 \times 10^{-19} N_{HI}/(T_s\, \Delta v)$ for a Gaussian profile from a uniform cloud of linewidth Δv (FWHM) in km s^{-1}. T_s is the excitation temperature of the transition, which is often, but not always, equal to the gas kinetic temperature (e.g., Liszt 2001). But the real interstellar medium is not uniform, and the typical 21cm profile consists of several blended components formed in regions of different temperature. If the line is optically thin at all velocities there is no dependence of N_{HI} on T_s and $N_{HI} = 1.8 \times 10^{18} \int T_b dv$ cm^{-2}. The optically thin assumption always gives a lower limit on N_{HI}.

In directions where part of the line has $\tau \geq 0.1$ the concept of a meaningful T_s becomes ambiguous and there is no unique solution for N_{HI} from 21cm emission data alone (e.g., Kalberla et al. 1985; DL90; Dickey 2002). An HI cloud at 100 K with $\Delta v = 10$ km s^{-1} has $\tau = 0.1$ for $N_{HI} = 2 \times 10^{20}$ cm^{-2}, so the 21cm line in an average direction (see Fig. 2) should be treated as if it has components which are not optically thin.

Digression: The Two-phase ISM. Theory tells us that under some conditions HI can exist in two stable phases at a single pressure: a warm phase whose temperature is thousands of Kelvins, and a cool phase whose temperature is ≤ 100 K (e.g., Field, Goldsmith, & Habing 1969; Wolfire et al. 2003). Observations suggest that reality is not so bimodal (e.g., Liszt 1983), but the

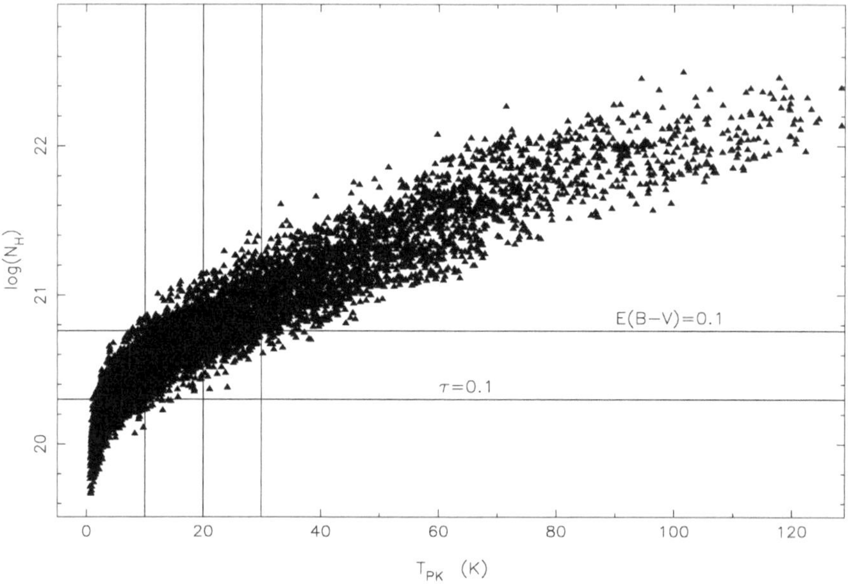

Figure 2. The total N_{HI} calculated for $T_s = 135$ K vs the peak line brightness temperature, Tpk, for about 10^4 21cm spectra from the LD survey. The sample includes data from $-90° \leq b \leq 90°$ every $5°$ in longitude between $30°$ and $180°$. Spectra above the $\tau = 0.1$ line likely contain components which have some opacity. Directions with HI spectra above the $E(B-V) = 0.1$ line likely intersect regions of some molecular gas.

generalization is still useful — the ISM does contain cool HI with a high 21cm line opacity and warm HI with a low opacity (e.g., Heiles & Troland 2003). In the Solar neighborhood there is more mass in the warm HI than the cold (Liszt, 1983; Dickey & Brinks, 1993). The cold phase fills a much smaller volume than the warm phase and has a smaller scale-height as well, so at high Galactic latitudes many sight lines skirt the clouds and intersect predominantly "intercloud" medium, which has a low opacity because of its high temperature and turbulence. In these directions N_{HI} can be determined quite well.

Deriving N_{HI} In Practice. A 21cm profile should always be evaluated for several excitation temperatures, e.g., $T_s = 1000$ K and 135 K. If the resulting values of N_{HI} differ significantly, where "significantly" depends on the accuracy one needs, it is likely that a thorough investigation of the direction of interest must be made by measuring $\tau(v)$ in absorption against nearby radio continuum sources (DL90; Dickey 2002). A sense of how often this might be required is given in Figure 2, where N_{HI} is plotted against the peak T_b in the line, T_{pk}, for a sample of 10^4 spectra from the LD survey. The line at 2×10^{20} cm^{-2} labeled $\tau = 0.1$ marks, very approximately, the N_{HI} below which the 21cm line is most likely optically thin. Conversely, Fig. 2 suggests that any profile with $T_{pk} > 10$ K should be tested for possible opacity. This applies to about 80% of the directions in the Figure.

4. Angular Structure

> The Mississippi [river] is remarkable in still another way — its disposition to make prodigious jumps by cutting through narrow necks of land, and thus straightening and shortening itself... In the space of 176 years the Lower Mississippi has shortened itself 242 miles. That is an average of a trifle over one mile and a third per year. Therefore, any calm person, who is not blind or idiotic, can see that in the Old Oolitic Silurian Period, just a million years ago next November, the Lower Mississippi River was upwards of 1,300,000 miles long, and stuck out over the Gulf of Mexico like a fishing-rod. And by the same token any person can see that 742 years from now the Lower Mississippi will be only a mile and three quarters long, and Cairo [Illinois] and New Orleans will have joined their streets together, and be plodding comfortably along under a single mayor and a mutual board of aldermen. There is something fascinating about science. One gets such wholesale returns of conjecture out of such a trifling investment of fact.
>
> — Mark Twain (1883), in *Life on the Mississippi*

Use of 21cm data to determine the ISM toward an extragalactic object often requires extrapolation over orders of magnitude in solid angle: from the area covered by the radio antenna beam, to the often infinitesimal area of the object under study. The highest angular resolution typically obtainable for a Galactic 21cm HI emission spectrum (with good sensitivity) is $\sim 1'$, and this is very much larger than the size of an AGN. Hence the need for extrapolation.

This is a vexing subject which causes some quite respectable scientists to loose their heads (I won't give explicit references, you know who you are). A few take the situation as license to rearrange the ISM into whatever preposterous filigree of structure simplifies their work — contriving an ISM which appears smooth on scales constrained by the data, but which goes crazy in finer detail. They ignore the wealth of data on the real small-scale structure of interstellar hydrogen. The situation has been further confused by the reported discovery of an anomalous population of tiny, dense HI clouds, whose significance, not to say reality, is now known to have been exaggerated. Twain's warning against thoughtless extrapolation is especially appropriate to this topic.

The Galactic ISM is not a free parameter! Is there structure in the total Galactic N_{HI} on all angular scales? Yes! Does this introduce errors when extrapolating to small angles? Yes. Are the errors important? Usually not! From point to point across the sky HI clouds come and go, and line components shift shape and velocity, but the dominant changes in *total* N_{HI} are usually on the largest linear scales, and do not cause large fractional fluctuations over small angles ($\leq 0.5°$).

Most structure in interstellar HI results from turbulence, characterized by a power-law spectrum with an exponent always less than -2; this has been determined experimentally and is understood theoretically (e.g., Green 1993; Lazarian & Pogosyan 2000; Deshpande, Dwarakanath, & Goss 2000; Dickey et al. 2001 and references therein). Small angles in nearby gas (e.g., at high Galactic latitude) correspond to small linear scales where there is relatively little structure. If the sight line intersects a distant high-velocity cloud then small angles may correspond to large spatial scales and the variations in that spectral component will be larger, but this is usually a problem for the *total* N_{HI} only in directions dominated by distant gas (see §6). Cold HI may have more structure than warm HI, and molecular clouds even more, but as a practical matter, the extrapolation to small angles introduces large errors only when a significant part of the gas in a particular direction is molecular or of anomalous origin, e.g., comes from a high-velocity cloud. Examination of 21cm spectra around the position of interest should give adequate warning of possible structure which then would require higher resolution observations to measure.

A lack of appreciation for the effects of power-law turbulence at small angles was one reason why the early, high-resolution, VLBI studies were interpreted as evidence for an anomalous population of extremely dense HI clouds with sizes of tens of AU. This in turn led some to assume that there must be extreme fluctuations in the *total* N_{HI} on very small angular scales. Somewhere Mark Train was chuckling. But with more complete observations (Faison 2002, Johnston et al. 2003) and a better understanding of how to interpret them (Deshpande 2000) the anomaly has disappeared almost entirely. The measured small-scale fluctuations in N_{HI} are likely to be entirely consistent with the known power laws (Deshpande 2000). Recent observations of 21cm absorption toward pulsars probing linear scales of 0.005–25 AU find no evidence of spatial structure at the 1σ level of $\Delta e^{-\tau} = 0.035$ (Minter, Balser & Karlteltepe 2003), and other careful observations toward pulsars suggest that some of the initial claims of small-angle fluctuations in HI absorption might be in error (Stanimirovic et al. 2003). The issue of small-scale structure in the ISM is interesting, and there are anomalous directions, e.g., toward 3C 138, where clumping in cold gas appears to be significant (Faison 2002), though here the total N_{HI} is $> 2 \times 10^{21}$ cm^{-2} and not a typical extragalactic sightline.

Recently Barnes & Nulsen (2003) combined interferometric and single-dish data to measure 21cm emission toward three high-latitude clusters of galaxies and found limits on fluctuations in N_{HI} on scales of $1'$–$10'$ of $< 3\%$ to $< 9\%$ (1σ). DL90 had suggested that on these angular scales $\sigma(N_{HI})/\langle N_{HI}\rangle \leq 10\%$, an estimate which has been controversial, but now appears somewhat conservative. I believe that, as concluded in DL90, directions without significant H_2, and without significant anomalous-velocity HI, are unlikely to contain small-scale angular structure in HI that is a major source of error in estimates of the effects of the Galaxy on extragalactic observations.

5. Molecular Hydrogen and Helium

Molecular Hydrogen. The 21cm data can be used to predict the likelihood that there is molecular gas along the line of sight, because there is usually cool HI associated with molecular clouds (though the converse may not necessarily be true, see Gibson 2002). Direct observations of H_2 show that $\geq 10\%$ of the neutral ISM is in molecular form when a sight line has a reddening $E(B-V) \geq 0.1$, equivalent to $N_H = 5.8 \times 10^{20}$ (Bohlin, Savage & Drake 1978; Rachford et al. 2003). This line is marked in Figure 2. About half of the directions in the Fig. 2 sample lie above the $E(B-V) = 0.1$ line. This suggests that for $T_{pk} \geq 20$ K there may be molecular gas somewhere along the path, an unfortunate circumstance, for an accurate N_{H_2} will then be quite difficult to obtain in the absence of bright UV targets in these directions.

Helium. Interstellar H_e^0 and H_e^+ both contribute to the opacity at $E > 13.6$ eV but their abundance cannot be determined by simply scaling N_{HI} and N_{H_2}, for a large fraction of the ISM is ionized (Reynolds et al. 1999). Near the Sun, the mass in ionized gas is about one-third the mass of HI, with substantial variations in different directions (Reynolds 1989). The fractional He ionization in the medium seems low (Reynolds & Tufte 1995). Maps of the sky in H_α show the location of the brighter ionized regions, but the H_α intensity is proportional to $n_p n_e$, not to N_{He}. The dispersion measure of pulsars gives N_e exactly, but there are not enough pulsars to map out the Galactic N_e to sufficient precision. Kappes, Kerp and Richter (2003) studied the X-ray absorbing properties of a large area of the high-latitude sky and conclude that 20–50% of the X-ray absorbing material is ionized and not traced by HI (see also Boulanger et al. 2001 and references therein). Unlike molecular gas, whose effects can be neglected in directions of low N_{HI}, the ionized component appears to cover the sky. Because we know so little about the detailed structure of the ionized component of the ISM, *it probably contributes the most significant uncertainties in our understanding of Galactic foregrounds at high Galactic latitude.*

6. The Chandra Deep Field North (CDFN)

The Chandra Deep Field North is an example of a direction which lacks dense gas, and thus one set of complexities, but has other features of interest (this direction is also the Hubble Deep Field, but as my comments are probably of more interest to X-ray than optical astronomers, I will keep the X-ray name). Figure 3 shows a 21cm spectrum from the LD survey at $35'$ resolution toward the CDFN ($\ell, b = 126° + 55°$). As a Galactic astronomer I find this sight-line

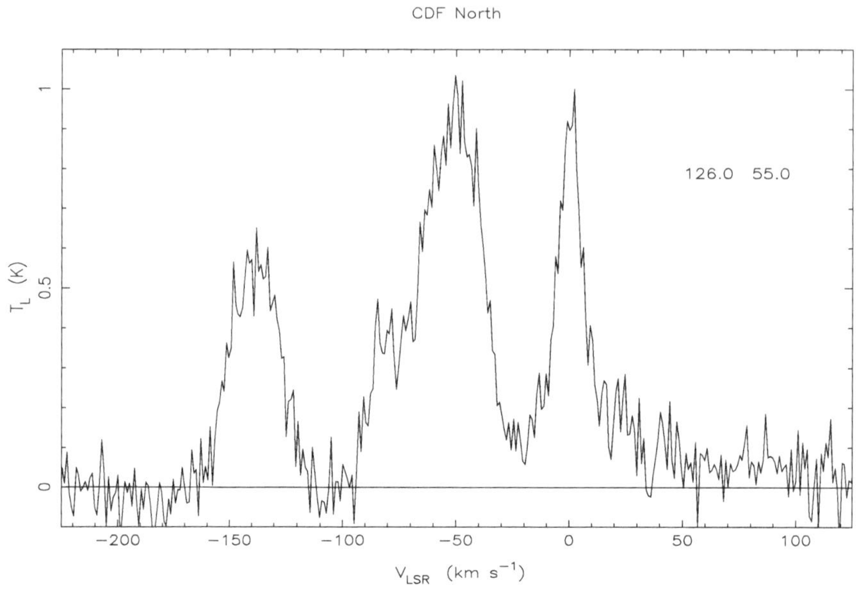

Figure 3. The 21cm spectrum toward the HDFN/CDFN at $35'$ resolution from the LD survey.

fascinating because it intersects a high-velocity cloud, an intermediate velocity cloud, and low-velocity "disk" gas, (from left to right in the spectrum), containing, respectively, 20%, 50% and 30% of the total N_{HI}. This gas is almost certainly optically thin in the 21cm line. The low-velocity gas further shows evidence of two components, one warm and one cool. It is uncommon to find a sight line where anomalous-velocity gas dominates the total N_{HI}, but that is the case for the CDFN and nearby areas of the sky.

Observations of the CDFN with the 100 meter Green Bank Telescope (GBT) of the NRAO at $9'$ resolution show a spectrum similar to Figure 3 at the low and intermediate velocities, but only $1/3$ as bright toward the high-velocity cloud. Figure 4 shows why: the high-velocity cloud has a gradient of nearly two orders of magnitude in N_{HI} near the CDFN. It becomes the brightest component in the entire spectrum to the west, contributing nearly 10^{20} cm^{-2} or more than 40% of the total column density, while to the east it is almost undetectable. This high degree of angular structure is typical of high-velocity clouds, which lie far beyond the local disk gas (Wakker & van Woerden 1997). The intermediate velocity gas over this field has a smaller dependence on angle, with a factor ~ 2 change in N_{HI} and most structure to the south and east.

In a direction (like the CDFN) where much of the hydrogen is in a high-velocity cloud, the total N_{HI} is likely to be a poor predictor of many interesting quantities like $S_{100\mu}$ and $E(B-V)$. For example, high-velocity clouds have a lower emissivity per H atom at 100μ than disk and intermediate-velocity gas

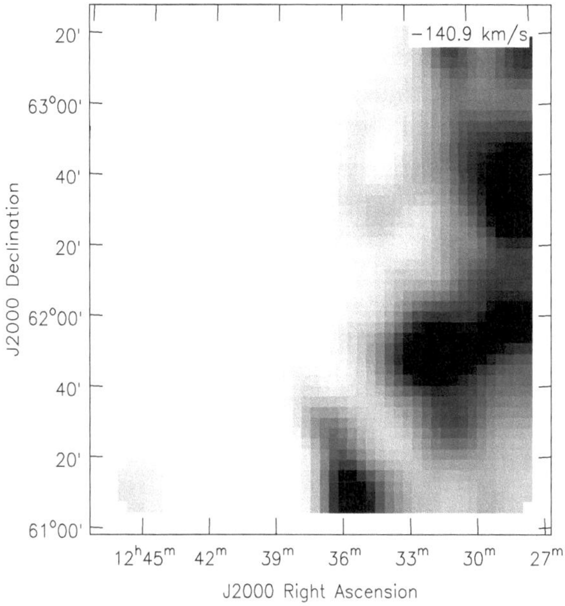

Figure 4. The structure in the high-velocity cloud at -140.9 km s^{-1} toward the Chandra Deep Field North as measured at $9'$ resolution with the Green Bank Telescope (Lockman 2004, in preparation). The emission line varies from $T_{pk} > 2$ K to $T_{pk} < 0.05$ K across the field. The CDFN is at $12^h36^m50^s + 62°13'00''$

because they lack of dust and/or have low heating (Wakker & Boulanger 1986). The particular high-velocity cloud that crosses the CDFN has an abundance of metals only about 10% that of solar (Richter et al. 2001). The intermediate velocity HI component also probably has different properties than disk gas at some wavelengths because of its different dynamical history. Thus, unlike most directions on the sky, where the kinematics of Galactic HI is irrelevant to its effects on extragalactic observations, the direction of the CDFN may be an exception, and require a specific HI component analysis. In contrast to the CDFN, preliminary GBT observations of the Chandra Deep Field South show that it has a HI spectrum dominated by a single, weak, low-velocity line. Quite boring compared to the CDFN.

7. Concluding Comments

The 21cm line can be a powerful tool for estimating the influence of the Galactic ISM on extragalactic observations, but it must be used with some thought, and it does not give the complete picture. Unresolved angular structure in the 21cm data is unlikely to dominate the error budget in most directions. For $N_{HI} > 6 \times 10^{20}$ cm^{-2} it is probable that there is some molecular hydrogen along the sight line and the total N_H will be difficult to determine. But in my opinion, it is the poorly-understood ionized component of the ISM which introduces the most serious uncertainties for directions with little molecular gas, and it affects observations in all directions. We are entering an era when highly accurate 21cm data will be available over the entire sky, and then, (though likely even now) the limits on understanding the influence of the ISM on extragalactic observations will lie not in the uncertainties in N_{HI}, but in N_{H_2} and N_{He}, and in the relationship between the dust and the gas. Every observation of an extragalactic object is an opportunity to learn something about the ISM. We should not let such opportunities go to waste.

Acknowledgements I thank J.M. Dickey, M. Elvis, C.E. Heiles, A.H. Minter, and R.J. Reynolds for comments on the manuscript.

References

Balucinska-Church, M., & McCammon, D. 1992, ApJ, 400, 699

Barnes, D.G., & Nulsen, P.E.J. 2003, MNRAS, 343, 315

Bohlin, R.C., Savage, B.D., & Drake, J.F. 1978, ApJ, 224, 132

Boulanger, F., Bernard, J-P., Lagache, G., & Stepnik, B., 2001, "The Extragalactic Infrared Background and its Cosmological Implications", IAU Symp. 204, ed. M. Harwit & M.G. Hauser, ASP, p. 47

Deshpande, A.A. 2000, MNRAS, 317, 199

Deshpande, A.A., Dwarakanath, K.S., & Goss, W.M. 2000, ApJ, 543, 227

Dickey, J.M., & Brinks, E., 1993, ApJ, 405, 153

Dickey, J.M., & Lockman, F.J., 1990, ARAA, 28, 215 (DL90)

Dickey, J.M., McClure-Griffiths, N.M., Stanimirovic, S., Gaensler, B.M., & Green, A.J. 2001, ApJ, 561, 264

Dickey, J.M. 2002, in "Seeing Through the Dust", ASP Conf. Ser. Vol. 276, ed. A.R. Taylor, T.L. Landecker, & A.G. Willis, p. 248

Elvis, M., Lockman, F.J., & Fassnacht, C. 1994, ApJS, 95, 413

Faison, M.D. 2002, in "Seeing Through the Dust", ASP Conf. Ser. Vol. 276, ed. A.R. Taylor, T.L. Landecker, & A.G. Willis, p. 324

Field, G.B., Goldsmith, D.W., & Habing, H.J. 1969, ApJ, 155, L149

Gibson, S.J. 2002, in "Seeing Through the Dust", ASP Conf. Ser. Vol. 276, ed. A.R. Taylor, T.L. Landecker, & A.G. Willis, p. 235

Green, D.A. 1993, MNRAS, 262, 328

Hartmann, D. & Burton, W.B. 1997, "Atlas of Galactic Neutral Hydrogen" (Cambridge Univ. Press) (The LD survey)

Hauser, M.G. 2001, "The Extragalactic Infrared Background and its Cosmological Implications", IAU Symp. 204, ed. M. Harwit & M.G. Hauser, ASP, p. 101

Heiles, C. & Troland, T.H. 2003, ApJ, 586, 1067

Jahoda, K., Lockman, F.J., & McCammon, D. 1990, ApJ, 354, 184

Johnston, S., Koribalski, B., Wilson, W., & Walker, M. 2003, MNRAS, 341, 941

Kalberla, P.M.W., Schwarz, U.J., & Goss, W.M. 1985, A&A, 144, 27
Kappes, M., Kerp, J., & Richter, P. 2003, A&A, 405, 607
Knee, L.B.G. 2002, in "Seeing Through the Dust", ASP Conf. Ser. Vol. 276, ed. A.R. Taylor, T.L. Landecker, & A.G. Willis, p. 50
Kulkarni, S.R., & Heiles, C. 1987, in "Interstellar Processes", ed. D.J. Hollenbach & H.A. Thronson, Jr., Reidel, p. 87
Lazarian, A., & Pogosyan, D. 2000, ApJ, 537, 720
Liszt, H.S. 1983, ApJ, 275, 163
Liszt, H.S. 2001, A&A, 371, 698
Lockman, F.J., Jahoda, K., & McCammon, D. 1986, ApJ, 302, 432
Lockman, F.J. 2002 in "Seeing Through the Dust", ASP Conf. Ser. Vol. 276, ed. A.R. Taylor, T.L. Landecker, & A.G. Willis, p. 107
McClure-Griffiths, N.M. 2002, in "Seeing Through the Dust", ASP Conf. Ser. Vol. 276, ed. A.R. Taylor, T.L. Landecker, & A.G. Willis, p. 58
Minter, A.H., Balser, D.S., & Karlteltepe, J. 2004, ApJ (in press)
Miville-Deschênes, M-A., Boulanger, F., Joncas, G., & Falgarone, E. 2002, A&A, 381, 209
Rachford, B.L., Snow, T.P., Tumlinson, J., Shull, J.M., et al. (2002), ApJ, 577, 221
Reynolds, R.J. 1989, ApJ, 339, L29
Reynolds, R.J., Haffner, L.M., & Tufte, S.L. 1999, in "New Perspectives on the Interstellar Medium", ASP Conf. Ser. Vol. 168, eds. A.R. Taylor, T.L. Landecker, & G. Joncas, p. 149
Reynolds, R.J., & Tufte, S.L. 1995, ApJ, 439, L17
Richter, P. et al. 2001, ApJ, 559, 318
Stanimirovic, S., Weisberg, J.M., Hedden, A., Devine, K.E., & Green, J.T. 2003, ApJL (in press; astro-ph/0310238)
Taylor, A.R. et al. 2002, in "Seeing Through the Dust", ASP Conf. Ser. Vol. 276, ed. A.R. Taylor, T.L. Landecker, & A.G. Willis, p. 68
Twain, M. 1883, *Life on the Mississippi*, James R. Osgood & Co.: Boston
Wakker, B.P., & Boulanger, F. 1986, A&A, 170, 84
Wakker, B.P., & van Woerden, H. 1997, ARAA, 35, 217
Wolfire, M.G., McKee, C.F., Hollenbach, D., & Tielens, A.G.G.M. 2003, ApJ, 587, 278

IV

HARD X-RAY EXCESSES AND NON-THERMAL PROCESSES

HARD X-RAY EXCESSES IN CLUSTERS OF GALAXIES AND THEIR NON-THERMAL ORIGIN

R.Fusco-Femiano
Istituto di Astrofisica Spaziale e Fisica Cosmica, CNR, Roma, Italy
dario@rm.iasf.cnr.it

Abstract The first part of the paper reports all the results obtained by *BeppoSAX* observations concerning the search for non-thermal components in the spectra of clusters of galaxies, while the origin of the hard X-ray excesses detected in Coma, A2256 and, at lower confidence level, in A754 is discussed in the second part.

1. Introduction

It is well known that in addition to the hot intracluster gas shown by X-ray measurements of the thermal bremsstrahlung emission in the energy range 1-10 keV there is now a compelling evidence for the existence of magnetic fields and relativistic particles in the intracluster medium (ICM) of some clusters of galaxies. The existence of these non-thermal elements is directly demonstrated by the presence of diffuse radio emission (radio halos and relics) detected in a growing fraction of the observed clusters. Furthermore, FR measurements toward discrete radio sources in clusters, coupled with the X-ray emission from the hot ICM, have indicated the presence of $\sim \mu G$ magnetic field strengths. The presence of non-thermal quantities seems to be reinforced by recent researches that have unveiled new spectral components in the ICM of some clusters of galaxies, namely a cluster soft excess discovered by *EUVE* (Lieu *et al.* 1966) and a hard X-ray excess (HXR) excess detected by *BeppoSAX* and *RXTE* .

Non-thermal HXR emission was predicted at the end of seventies in clusters of galaxies showing extended radio emission, radio halos or relics (see Rephaeli 1979) since the same radio synchrotron electrons can interact with the CMB photons to give inverse Compton (IC) non-thermal X-ray radiation. Attempts to detect non-thermal emission from a few clusters of galaxies were performed with various experiments : balloon experiments, *HEAO-1*, the OSSE experiment onboard the *Compton-GRO* satellite and by joint analysis of *RXTE* & *ASCA* data (Bazzano *et al.* 1984,90; Rephaeli, Gruber & Rothschild

R. Lieu and J. Mittaz (eds.), Soft X-Ray Emission from Clusters of Galaxies and Related Phenomena, 125–135.

1987; Rephaeli & Gruber 1988; Rephaeli, Ulmer & Grubber 1994; Delzer & Henriksen 1998), but all these experiments reported essentially flux upper limits. The search for non-thermal emission was one of the central scientific subjects carried out by *BeppoSAX* , starting from 1997, exploiting the unique spectral capabilities of the Phoswich Detector System (PDS) able to detect HXR radiation in the 15-200 keV energy range (Frontera *et al.* 1997).

2. BeppoSAX observations

BeppoSAX observed seven clusters of galaxies with the main objective to detect non-thermal components in their X-ray spectra.

2.1 Coma cluster

The first cluster was Coma observed in December 1997 for an exposure time of about 91 ksec (Fusco-Femiano *et al.* 1999). A nonthermal excess with respect to the thermal emission was observed at a confidence level of about 4.5σ (see Fig. 1). The χ^2 value has a significant decrement when a second component, a power law, is added. On the other hand, if we consider a second thermal component, instead of the non-thermal one, the best-fit requires a temperature of $\sim$150 keV (>20 keV at 90% and >10 keV at 99%). So, this unrealistic value may be interpreted as a strong indication that the detected hard excess is due to a non-thermal mechanism. In the same time a *RXTE* observation of the Coma cluster (Rephaeli, Gruber & Blanco 1999) showed evidence for the presence of a second component in the spectrum of this cluster. In particular the authors argued that this component is more likely to be non-thermal , rather than a second thermal component at lower temperature.

2.2 A2199

The cluster was observed in April 1997 for $\sim$100 ksec (Kaastra *et al.* 1999). The MECS data in the range 8-10 keV seem to show the presence of a hard excess with respect to the thermal emission at a confidence level of $\sim 3.3\sigma$. The PDS data are instead not sufficient to prove the existence of a hard tail. We have re-analyzed the MECS data finding that only a point is above the thermal model at the level of $\sim 2\sigma$ (Fusco-Femiano *et al.* 2002). However, the cluster is planned to be observed by *XMM-Newton* that should be able to discriminate between these two different results of the MECS data analysis, considering the low average gas temperature of about 4.5 keV (David *et al.* 1993).

2.3 A2256

The cluster A2256 is the second cluster where *BeppoSAX* detected a clear excess (Fusco-Femiano *et al.* 2000) at about 4.6σ above the thermal emission

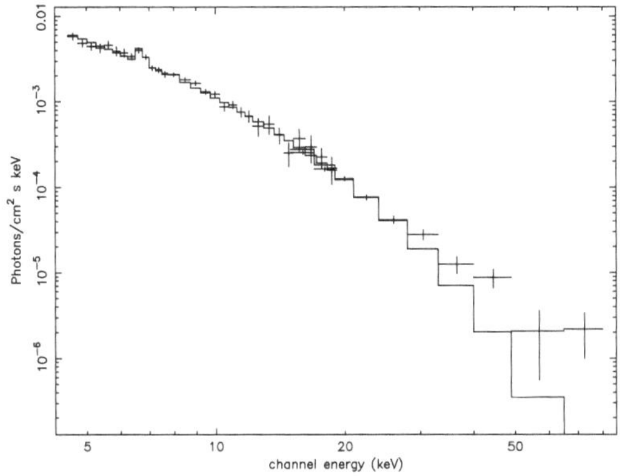

Figure 1. Coma cluster - HPGSPC and PDS data. The continuous line represents the best fit with a thermal component at the average cluster gas temperature of $8.5^{+0.6}_{-0.5}$ keV.

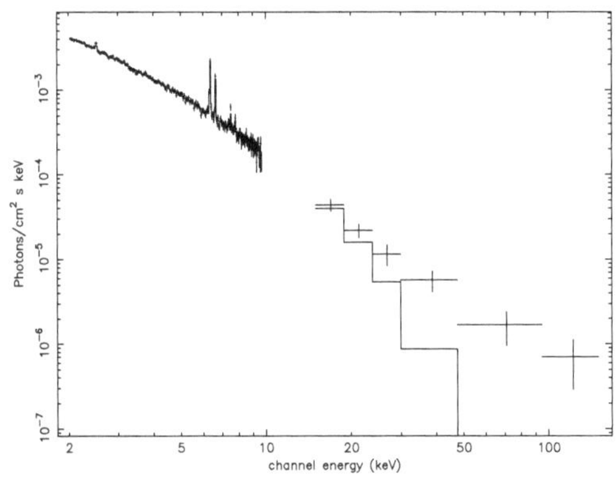

Figure 2. Abell 2256 - MECS and PDS data. The continuous line represents the best fit with a thermal component at the average cluster gas temperature of 7.47±0.35 keV.

(see Fig. 2). Also in this case the χ^2 value has a significant decrement when a second component, a power law, is added and also in this case the fit with a second thermal component gives an unrealistic temperature of ~200 keV (>25 kev at 90% and >13 keV at 99%) which can be interpreted as evidence in favour of a non-thermal mechanism for the second component present in the X-ray spectrum of A2256.

2.4 A1367

A *BeppoSAX* observation of Abell 1367 has not detected hard X-ray emission in the PDS energy range above 15 keV (P.I.: Rephaeli). A1367 is a near cluster (z=0.0215) that shows a relic at a distance of about 22′ from the center and a low gas temperature of ~ 3.7 keV (David *et al.* 1993) that might explain lack of thermal emission at energies above 15 keV.

2.5 A3667

A3667 is one of the most spectacular clusters of galaxies . It contains one of the largest radio sources in the southern sky with a total extent of about 30′ which corresponds to about $2.6h_{50}^{-1}$ Mpc. A long observation with the PDS (effective exposure time 44+69 ksec) reports a clear detection of hard X-ray emission up to about 35 keV at a confidence level of $\sim 10\sigma$. Instead, the fit with a thermal component at the average gas temperature indicates a marginal presence of a non-thermal component. Given the presence of such a large radio region in the NW of the cluster, a robust detection of a non-thermal component might be expected instead of the upper limit reported by *BeppoSAX* . One possible explanation may be related to the radio spectral structure of the NW relic (Fusco-Femiano *et al.* 2001).

2.6 A754

The rich and hot cluster A754 is considered the prototype of a merging cluster. X-ray observations report a violent merger event in this cluster (Henry & Briel 1995; Henriksen & Markevitch 1996; De Grandi & Molendi 2001), probably a very recent merger as shown by a numerical hydro/N-body model (Roettiger, Stone, & Mushotzky 1998). A long *BeppoSAX* observation of A754 shows a non-thermal excess at energies above about 45 keV with respect to the thermal emission at the temperature of 9.4 keV (Fusco-Femiano *et al.* 2003). The excess is at a level of confidence of 3.2σ (see Fig. 3).

2.7 A119

Finally, the last cluster observed by *BeppoSAX* to detect non-thermal component was A119. The X-ray observations have shown a rather irregular and asymmetric X-ray brightness suggesting that the cluster is not completely relaxed and may have undergone a recent merger (Cirimele *et al.* 1997; Markevitch *et al.* 1998; Irwin, Bregman & Evrard 1999; De Grandi & Molendi 2001). The PDS reports an excess with respect to the thermal emission at the average gas temperature measured by the MECS (5.66 ± 0.16 keV) at a confidence level of $\sim 2.8\sigma$. A119 does not show evidence of a radio halo or relic, but the presence of a recent merger event could accelerate particles able to emit nonther-

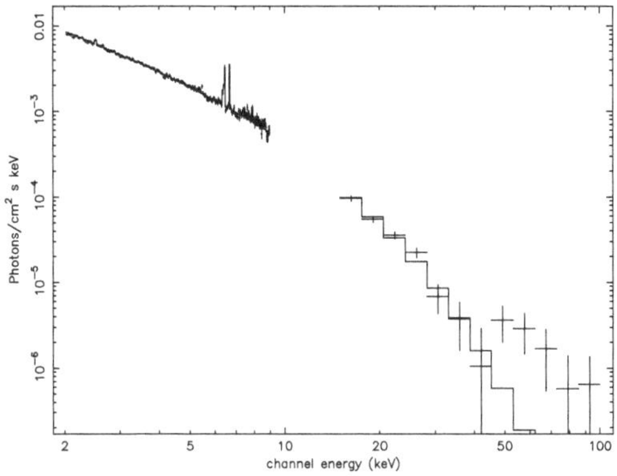

Figure 3. Abell 754 - MECS and PDS data. The continuous lines represent the best fit with a thermal component at the average cluster gas temperature of $9.42^{+0.16}_{-0.17}$ keV

mal emission in the PDS energy range. However, the presence of 7 QSO in the field of view of the PDS with redshift in the range 0.14-0.58 makes very unlikely that this possible excess may be due to a diffuse source (Fusco-Femiano *et al.* 2002).

3. Possible interpretations of the observed non-thermal HXR excesses in Coma, A2256 and A754

The first possible explanation for the detected excesses is emission by a point source in the field of view of the PDS. It has been verified that sources like X Comae or the three quasars reported by *XMM-Newton* in Coma and the QSO in A2256 cannot be responsible for the detected excesses (Fusco-Femiano *et al.* 1999; 2002), but it is not possible to exclude that an obscured source, like Circinus (Matt *et al.* 1999) a Seyfert 2 galaxy very active at high X-ray energies, may be present in the field of view of the PDS able to simulate the observed non-thermal emission. With the MECS image it is possible to exclude the presence of this kind of source only in the central region of about $30'$ in radius unless the obscured source is within $2'$ of the central bright core. The probability to find an obscured source in the field of view of the PDS is of the order of 10% (Fusco-Femiano *et al.* 2002). In the case of A754 the radio galaxy 26W20 (Harris *et al.* (1980) is present in the field of view of the PDS that shows a X-ray bright core similar to that of a BL Lac object. The fit with a Synchrotron Self-Compton (SSC) model to the SED requires a flat energy index of about 0.3 to extrapolate the flux detected by *ROSAT* in the PDS energy range. The inclusion of the PDS points makes difficult to fit well all the points of the SED (Fusco-Femiano *et al.* 2003).

A support against the point source interpretation is given by the spectrum of A2256 re-observed after two years from the previous one. The two spectra are consistent (Fusco-Femiano *et al.* 2002) and both observations comprise two exposures with a time interval of $\sim$ 1 year and $\sim$ 1 month, respectively and all these observations do not show significant flux variations. Besides, a second recent *RXTE* observation of the Coma cluster (Rephaeli & Gruber 2002) confirms the previous one (Rephaeli, Gruber, & Blanco 1999) of a likely presence of a non-thermal component in the spectrum of the cluster. These results strongly support the idea that a diffuse non-thermal mechanism involving the ICM is responsible for the observed excesses, also considering that Coma, A2256 and A754 present extended radio regions.

Another interpretation is that the non-thermal emission is due to relativistic electrons scattering the CMB photons and in particular the same electrons responsible for the extended radio regions present in these clusters. In particular, in the case of the Coma cluster we derive a volume-averaged intracluster magnetic field, B_X, of 0.15 μG, using only observables, combining the X-ray and radio data. The value of the magnetic field derived by this *BeppoSAX* observation seems to be inconsistent with the measurements of Faraday rotation (FR) of polarized radiation of sources through the hot ICM that give a line-of-sight value of B_{FR} of the order of 2-6 μG (Kim *et al.* 1990; Feretti *et al.* 1995). Regarding this discrepancy it has been shown that IC models that include the effects of more realistic electron spectra, combined with the expected spatial profiles of the magnetic fields, and anisotropies in the pitch angle distribution of the electrons allow higher values of the intracluster magnetic field , in better agreement with the FR measurements (Goldshmidt & Rephaeli 1993; Brunetti *et al.* 2001; Petrosian 2001). One of these models is the *two-phase* model of Brunetti *et al.* (2001) in which relativistic electrons injected in the Coma cluster by some processes (starbursts, AGNs, shocks, turbulence) during a first phase are (systematically) re-accelerated during a second phase for a relatively long time ($\sim$ 1 Gyr) up to present time. This model is able to explain the radio properties of the Coma cluster : the spectral cutoff and the radial spectral steepening, and the HXR excess present in this cluster. In this model the maximum energy of the re-accelerated electrons with Fermi II-like processes in the ICM is lower than 10^5 and consequently a cutoff in the electron spectrum might be present in the radio band ($\gamma_{Syn} \sim 10^4$). Considering that the energy of the IC electrons is lower than that of the Syn electrons for $B < 1\mu G$ ($\gamma_{syn}/\gamma_{HXR} \sim 3.5 \times B_{\mu G}^{-1/2}$), a cutoff close to γ_{syn} would reduce the synchrotron emission without affecting the IC in the HXR band. So, for a given ratio of IC/Syn emission the magnetic field must be higher to obtain the same synchrotron emission. This cutoff in the spectrum of the emitting electrons is confirmed by the spectral cutoff observed in Coma (Deiss *et al.* 1997).

The other radio halo property of Coma is the spectral steepening with the distance from the center (Giovannini *et al.* 1993) that can be explained if the magnetic field strength decreases with the distance from the center. In this case the corresponding frequency of the cutoff in the synchrotron spectrum ($\nu_c \sim B\gamma^2$) should decrease with the distance from the center yielding a possible radial steepening. The fit to the overall radio synchrotron properties and to the HXR excess provides a set of possible radial profiles of the cluster magnetic field. In particular, in the cluster core region where the RM observations are effective B is between 0.8-2 μG not much distant from the values inferred from RM measurements. Besides, recently, Newman, Newman, & Rephaeli (2002) have pointed out that many and large uncertainties are associated with the values of the magnetic field determined through FR measurements (see also Govoni *et al.* 2001).

However, alternative interpretations to the IC model have been proposed, some of these essentially motivated by the discrepancy between the values of B_X and B_{FR}. It has been proposed that a different mechanism may be given by non-thermal bremsstrahlung (NTB) from suprathermal electrons formed through the current acceleration of the thermal gas (Ensslin, Lieu, & Biermann 1999; Dogiel 2000; Sarazin & Kempner 2000; Blasi 2000; Liang, Dogiel, & Birkinshaw 2002).

Considering also the difficulties with the NTB model (Petrosian 2001; 2002) the presence of non-thermal phenomena in clusters of galaxies (diffuse radio emission, HXR excesses and probably also EUV emission) is usually explained in the framework of the IC model in which the alternative between primary and secondary electron populations represents the basis of the more recent theoretical works developed on the argument.

3.1 Primary electrons

Primary electrons can be injected in the ICM by different processes (Brunetti 2002) : one possibility is given by acceleration by shocks, in particular merger shocks can release a part of their energy into particle acceleration (Takizawa & Naito 2000; Miniati *et al.* 2001; Fujita & Sarazin 2001). In this case the typical extension of ~2 Mpc of radio halos requires an unreasonably high Mach number greater than 5 considering the short lifetime of the electrons. Furthermore, this implies a strong field compression with the consequence that the emitted synchrotron radiation is highly polarized, as in the case of radio relics, but in contrast with the very low polarization found in radio halos. Another possibility is given by re-accelerated electrons as shown by the *two-phases* model of Brunetti *et al.* (2001). A number of sources of relativistic electrons in galaxy clusters (radio galaxies, merger shocks, supernovae and galactic winds) can efficiently inject relativistic electrons in the cluster volume over cosmological time (e.g.: Sarazin 2002). Relativistic electrons with $\gamma \sim$ 100-300 can survive for some billion year and thus can be accumulated in the cluster volume. These electrons could be responsible for the EUV emission. To produce radio and HXR emission this population of electrons must be re-accelerated and a possible mechanism is given by a significant level of turbulence in the ICM produced by cluster mergers (Brunetti *et al.* 2001).

3.2 Secondary electrons

It is well known that a secondary electron population is invoked by the difficulty in explaining the extended radio halos by combining their $\sim$Mpc size, and the short radiative lifetime of the radio emitting electrons. To resolve this problem Dennison (1980) first pointed out that a possible source of relativistic electrons in radio halos is the continuous production of secondary electron due to the decay of charged pions generated in cosmic ray collisions in the ICM. This possibility has been reconsidered in detail by various authors (Blasi & Colafrancesco 1999; Dolag & Ensslin (2000); Ensslin 1999; Miniati *et al.* 2001)

3.3 Primary or secondary electrons ?

Some observational constraints seem to be able to discriminate between primary and secondary electrons (Brunetti 2002) :

a) The comparison between radio and soft X-ray brightness of a number of radio halos (Govoni *et al.* 2001; Feretti *et al.* 2001) indicates that the profile of the radio emission is broader than that of the X-ray thermal emission. This appears difficult to be explained within secondary models which would yield narrower radio profiles. This is due to the fact that the timescale of the p-p collision is inversely proportional to the gas density : $\propto n_{th}^{-1}$. As a consequence, for a given number density of the relativistic protons, the secondary electrons are expected to be injected in the denser regions and the radio emission would be stronger in the cluster core. A possibility to skip this problem is to admit an *ad hoc* increasing fraction of energy density of the relativistic protons with radius. This would imply at least in some cases an energetics of the relativistic protons higher than that of the thermal pool.

b) The spectral cutoff observed in Coma strongly point to the presence of a cutoff in the spectrum of the emitting electrons and this cutoff may be naturally accounted for if the synchrotron emission is produced by re-accelerated electrons (Brunetti *et al.* 2001). Instead, this cutoff is not naturally explained by secondary electrons unless to assume a cutoff in the energy distribution of the primary relativistic protons. This should be at $E_p < 50$ GeV to have a cutoff of the spectrum from the secondary electrons at $\sim$GHz frequencies. This seems to be in contrast with the observations of the spectrum of Galactic cosmic rays and with the theoretical expectations from the most accepted acceleration mechanisms.

In conclusion the radio spectral properties cannot be satisfied in a natural way in the case of secondary electrons. Of course, it is important to understand if the the synchrotron spectral properties of Coma are common among radio halos.

A confirm of the scenario described by the re-acceleration models (e.g. the two-phases model of Brunetti *et al.*) seems to be given by the radio and HXR observations of A754 (Fusco-Femiano *et al.* 2003). One possible explanation of the hard excess detected by *BeppoSAX* is tied to presence of diffuse radio emission recently discovered by Kassim *et al.* (2001) and confirmed by a deeper VLA observation at 1.4 GHz (Fusco-Femiano *et al.* 2003). The detection of the hard excess in A754 determines, in the framework of the IC model, a volume-averaged intracluster magnetic field of the same order ($B_X \sim 0.1\mu G$) of that determined in the Coma cluster (Fusco-Femiano *et al.* 1999). This value of B_X implies relativistic electrons at energies $\gamma \sim 10^4$ to explain the observed diffuse synchrotron emission. At these energies IC losses determine a cut-off in the electron spectrum as confirmed by the spectral cut-off observed in A754 obtained combining the VLA radio observation at 1.4 GHz with the observation of Kassim *et al.* (2001) at lower frequencies (a spectral index of 1.1 in the band 74-330 MHZ and 1.5 in the higher 330-1400 MHz frequency band). This spectral cutoff is similar to that reported in Coma. Considering that a cutoff in the radio emitting electrons cannot be explained in a natural way by a secondary electron population, as discussed above, the radio and HXR results obtained for A754 seem to reinforce the scenario that primary and not secondary electrons are responsible for non-thermal emission in clusters of galaxies.

4. Conclusions

BeppoSAX observed a clear evidence of non-thermal X-ray emission in two clusters, Coma and A2256, both showing extended radio regions. All the observations do not show variability in the non-thermal flux supporting the idea of a diffuse non-thermal mechanism involving the ICM. At a lower confidence level non-thermal HXR emission has been observed in A754. The detected excess in A754, if confirmed by a a deeper observation with imaging instruments, support the scenario that primary re-accelerated electrons are responsible for non-thermal phenomena in the ICM and reinforce the link between Mpc-scale radio emission and very recent or current merger processes.

BeppoSAX , as it is well known, has ceased its activity. The next missions able to search for non-thermal components are : *INTEGRAL* , launched in October, *ASTRO-E* , *CONSTELLATION* and *NEXT* with a great improvement in sensitivity. These missions will be operative in the next years, but the energy range and the spectral capabilities of *XMM-Newton* /EPIC give the possibility to localize non-thermal components in regions of low gas temperature ($\sim$ 4-5 keV). So with *XMM-Newton* we should have the possibility, by comparing the X-ray and radio structures, to constrain the profiles of the magnetic field and of relativistic electrons.

References

Bazzano, A., Fusco-Femiano, R., La Padula, C., Polcaro, V.F., Ubertini, P., & Manchanda, R.K. 1984, ApJ, 279, 515

Bazzano, A., Fusco-Femiano, R., Ubertini, P., Perotti, F., Quadrini, E., Court, A.J., Dean, N.A., Dipper, A.J., Lewis, R., & Stephen J.B. 1990, ApJ, 362, L51

Blasi, P., & Colafrancesco, S. 1999, APh, 12, 169

Blasi, P. 2000, ApJ, 532, L9

Brunetti, G., Setti, G., Feretti, L., & Giovannini, G., 2001, MNRAS, 320, 365

Brunetti, G. 2002, to appear in the Proc. of "Matter and Energy in Clusters of Galaxies", April 23-27 2002, Taiwan, S.Bowyer and C.-Y. Hwang, Eds; astro-ph/0208074

Cirimele. G., Nesci, R., & Trevese, D. 1997, ApJ, 475, 11

David, L.P., Slyz, A., Jones, C., Forman, W., & Vrtilek, S.D. 1993, ApJ, 412, 479

De Grandi, S., & Molendi, S. 2001, ApJ, 551, 153

Deiss, B.M., Reich, W., Lesch, H., & Nielebinski, R. 1997, A&A, 321, 55

Delzer, C., & Henriksen, M. 1998, AAS, 193, 3806

Dogiel, V.A. 2000, A&A, 357, 66

Dolag, K., Ensslin, T.A. 2000, A&A, 362, 151

Ensslin, T., Lieu, R., & Biermann, P.L. 1999, A&A, 344, 409

Ensslin, T.A. 1999, in Diffuse Thermal and Relativistic Plasma in Galaxy Clusters, eds : H.Boehringer, L.Feretti, P.Schuecker, MPE Report 271, 249

Feretti, L., Dallacasa, D., Giovannini, G., & Tagliani, A. 1995, A&A, 302, 680

Frontera, F., Costa, E., Dal Fiume, D., Feroci, M., Nicastro, L., Orlandini, M., Palazzi, E., & Zavattini, G. 1997, A&AS, 122, 357

Fujita, Y., & Sarazin, C.L. 2001, ApJ, 563, 660

Fusco-Femiano, R., Dal Fiume, D., Feretti, L., Giovannini, G., Grandi, P., Matt, G., Molendi, S., & Santangelo, A. 1999, ApJ, 513, L21

Fusco-Femiano, R., Dal Fiume, D., De Grandi, S., Feretti, L., Giovannini, G., Grandi, P., Malizia, A., Matt, G., & Molendi, S. 2000, ApJ, 534, L7

Fusco-Femiano, R., Dal Fiume, D., Orlandini, M., Brunetti, G., Feretti, L., & Giovannini, G. 2001, ApJ, 552, L97

Fusco-Femiano, R., Orlandini, M., De Grandi, S., Feretti, L., Giovannini, G., Bacchi, M., & Govoni, F. 2003, A&A, 398, 441

Fusco-Femiano *et al.* 2002, to appear in the Proc. of "Matter and Energy in Clusters of Galaxies", April 23-27 2002, Taiwan, S.Bowyer and C.-Y. Hwang, Eds; astro-ph/0207241

Giovannini, G., Feretti, L., Venturi, T., Kim, K.T., & Kronberg, P.P. 1993, ApJ, 406, 399

Goldshmidt, O., & Rephaeli, Y. 1993, ApJ, 411, 518

Govoni, F., Ensslin, T.A., Feretti, L., & Giovannini, G. 2001, A&A, 369, 441

Harris, D.E. *et al.* 1980, A&A, 90, 283

Henry, J.P., & Briel, U.G. 1995, ApJ, 443, L9

Henriksen, M.J., & Markevitch, M.L. 1996, ApJ, 466, L79

Irwin, J.A., Bregman, J.N., & Evrard, A.E. 1999, ApJ, 519, 518

Liang, H., Dogiel, V.A., & Birkinshaw, M. 2002, MNRAS, 337, 567

Lieu, R., Mittaz, J.P.D., Bowyer, S., Lockman, F.J., Hwang, C. -Y., & Schmitt, J.H.M.M. 1996, ApJ, 458, L5

Kaastra, J.S., Lieu, R., Mittaz, J.P.D., Bleeker, J.A.M., Mewe, R., Colafrancesco, S., & Lockman, F.J. 1999, ApJ, 519, L119

Kassim, N.E., Clarke, T.E., Enßlin, T.A., Cohen, A.S., & Neumann, D.M. 2001, ApJ, 559, 785

Kim, K.T., Kronberg, P.P., Dewdney, P.E., & Landecker, T.L. 1990, ApJ, 355, 29

Matt, G. *et al.* 1999, A&A, 341, L39

Markevitch, M., Forman, W.R., Sarazin, C.L., & Vikhlinin, A. 1998, ApJ, 503, 77

Miniati, F., Jones, W., Kang, H., & Ryu, D. 2001, ApJ, 562, 233

Newman, W.I., Newman, A.L., & Rephaeli, Y. 2002, astro-ph/0204451

Petrosian, V. 2001, ApJ, 557, 560

Petrosian, V. 2002, to appear in the Proc. of "Matter and Energy in Clusters of Galaxies", April 23-27 2002, Taiwan, S.Bowyer and C.-Y. Hwang, Eds; astro-ph/0207481

Rephaeli, Y. 1979, ApJ, 227, 364

Rephaeli, Y., Gruber, D.E., & Rothschild, R.E. 1987, ApJ, 320, 139

Rephaeli, Y., & Gruber, D.E. 1988, ApJ, 333, 133

Rephaeli, Y., Ulmer, M., & Gruber, D.E. & 1994, ApJ, 429, 554

Rephaeli, Y., Gruber, D.E., & Blanco, P. 1999, ApJ, 511, L21

Rephaeli, Y., & Gruber, D. 2002, ApJ, in press; astro-ph/0207443

Roettiger, K., Stone, J.M., & Mushotzky, R.F. 1998, ApJ, 493, 62

Sarazin, C.L., & Kempner, J.C. 2000, ApJ, 533, 73

Sarazin, C.L. 2002, in *Merging Processes of Galaxy Clusters*, eds : L.Feretti, I.M.Gioia, G.Giovannini, ASSL, Kluwer AC Publish, p. 1

Takizawa, M., & Naito, T. 2000, ApJ, 535, 586

THERMAL AND NON-THERMAL SZ EFFECT IN GALAXY CLUSTERS

Sergio Colafrancesco
INAF- Osservatorio Astronomico di Roma
Via Frascati 33, I-00040 Monteporzio (Roma), Italy
cola@mporzio.astro.it

Abstract We discuss the properties of the non-thermal SZ effect in galaxy clusters in comparison with the well know thermal SZ effect. Following a generalized approach to the derivation of the SZ effect, we show that it is possible to set interesting constraints to the presence and to the physical properties of additional populations of electrons (both of non-thermal and thermal origin) residing in the atmospheres of galaxy clusters.

Keywords: Cosmology, Galaxy clusters, CMB, Cosmic-ray interaction

1. The SZ effect in galaxy clusters: an introduction

Compton scattering of the Cosmic Microwave Background (CMB) radiation by hot Intra Cluster (hereafter IC) electrons – the Sunyaev Zel'dovich effect (hereafter SZE, Sunyaev & Zel'dovich 1972, 1980) – is a relatively simple process whose spectral and spatial imprints on the CMB radiation serves as indispensable cosmological and astrophysical probes (see Birkinshaw 1999 for a review). Such a scattering produces a systematic shift of the CMB photons from the Rayleigh-Jeans (RJ) to the Wien side of the spectrum.

Three main scattering processes can be considered in this framework:
i) the thermal SZE, i.e., the spectral distortion produced by the scattering of CMB photons by the thermal, hot ($T_e \sim 10^7 - 10^8$ K) IC gas electrons;
ii) the kinematic SZE, i.e., the spectral distortion produced by the bulk motion of the IC gas along the line of sight;
iii) the non-thermal SZE, i.e., the up-scattering of CMB photons by non-thermal relativistic electrons residing in the cluster atmosphere.

These processes produce different spectral distortion of the CMB spectrum which depend on the properties of the electron population which is responsible for the Compton scattering of the CMB photons (see Figs.1-2).

An approximate description of the scattering of an isotropic Planckian radiation

R. Lieu and J. Mittaz (eds.), Soft X-Ray Emission from Clusters of Galaxies and Related Phenomena, 137–146.

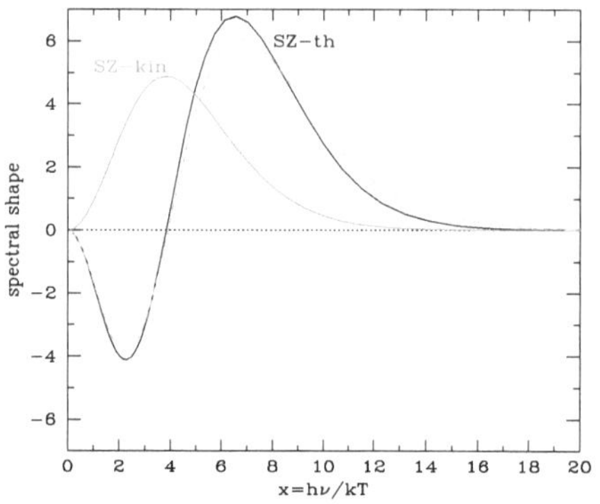

Figure 1. The spectral shape of the thermal SZE in the non-relativistic limit (blue solid curve) and including the relativistic corrections (dashed curve) and of the kinematic SZE (green solid curve) are shown for a Coma-like cluster.

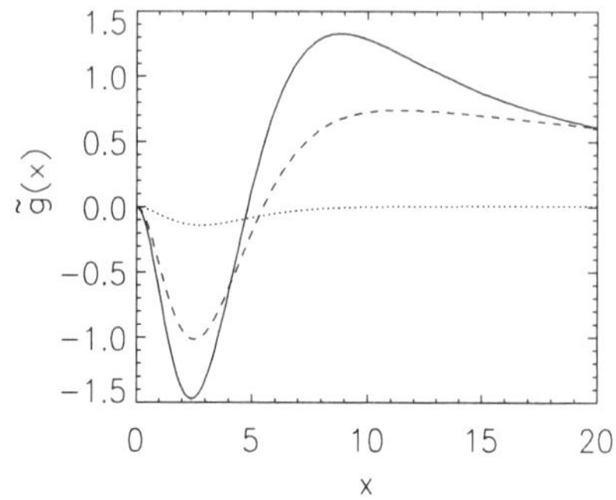

Figure 2. The function $\tilde{g}(x)$ for a non-thermal population with a power-law spectrum with $p_1 = 0.5$ (solid line), 1 (dashed) and 10 (dotted).

field by a non-relativistic Maxwellian electron population can be obtained by means of the solution of the Kompaneets equation. The resulting change in the spectral intensity, ΔI_{th} can be written as

$$\Delta I_{th} = 2\frac{(k_B T_0)^3}{(hc)^2} y_{th} g(x) , \tag{1}$$

where $x = h\nu/k_B T_0$ is the a-dimensional frequency, and the spectral shape of the thermal SZE is contained in the function

$$g(x) = \frac{x^4 e^x}{(e^x - 1)^2}\left[x\frac{e^x + 1}{e^x - 1} - 4\right] \tag{2}$$

which is zero at the frequency $x_0 = 3.83$ (or $\nu = 217$ GHz for a value of the CMB temperature $T_0 = 2.726$ K), negative at $x < x_0$ (in the RJ side) and positive at $x > x_0$ (in the Wien side). The Comptonization parameter is proportional to the integral along the line of sight ℓ of the kinetic pressure, $P_{th} = n_e k_B T_e$, of the IC gas and writes as

$$y_{th} = \frac{\sigma_T}{m_e c^2} \int d\ell \; P_{th} , \tag{3}$$

where n_e and T_e are the electron density and temperature of the IC gas, respectively, σ_T is the Thomson cross section, appropriate in the limit $T_e \gg T_0$, k_B is the Boltzmann constant and $m_e c^2$ is the rest mass energy of the electron. The

previous description of the thermal SZE is obtained in the non-relativistic limit and in the single scattering regime of the photon redistribution function. As such, it only provides an approximation of the SZE in galaxy clusters for low temperatures ($k_B T_e \lesssim 3$ keV) and low optical depth ($\tau \lesssim 10^{-3}$).
For high-T clusters, the calculation of the thermal SZE requires to take into account the appropriate relativistic corrections (see Colafrancesco et al. 2003 and references therein).
Other approximation widely used in the derivation of the SZE (the single-scattering limit and the presence of only one electron population residing in the cluster atmosphere) are not tenable within realistic physical conditions of many cluster atmospheres. We discussed in Colafrancesco et al. (2003) a more general approach to the derivation of the SZE in galaxy clusters.

1.1 Cosmology with the thermal SZE

The temperature decrement $\Delta T/T_0$ produced by the SZE in the direction of a galaxy cluster is independent of the redshift, as can be inferred from eq.(1). This unique feature of the SZE makes it a potentially powerful tool for investigating the high-z universe. In fact, the possible high-z clusters will be found by SZ surveys but missed in even the deepest X-ray planned observations. Future SZE surveys with optimized selection functions will thus provide ideal cluster samples for cosmological purposes (see Carlstrom et al. 2002 for a review). Not only complete catalogs of distant clusters will be available for studying the large-scale structure of the universe by using the standard statistical methods, but the SZE will also provide a direct view of the high-z universe allowing a more reliable extraction of the cosmological parameters. These cosmological milestones can be derived both by increasing the precision of the traditional SZE applications (e.g., cluster distance measurement, Hubble constant, cluster gas-mass fraction and peculiar velocities) and by exploiting the ability of deep and large SZE-cluster surveys to cleanly determine the evolution of the cluster abundance. In fact, the cluster z-distribution is critically sensitive to Ω_m and to the properties of the Dark Energy (or Ω_Λ) at sufficiently high redshift; moreover, sufficiently large and deep SZE surveys can efficiently constrain the equation of state of the Dark Energy.
However, to accomplish these cosmological tasks the detailed astrophysical study of the SZE is required. The detailed study of the SZE is furthermore motivated by the evidence for the existence of various electronic populations residing in the cluster atmosphere in addition to the hot IC gas.

2. Non-Thermal phenomena in galaxy clusters

In addition to the thermal IC gas, many galaxy clusters contain a population of relativistic electrons which produce a diffuse radio emission (radio halos and/or relics) via synchrotron radiation in a magnetized IC medium. The electrons which are responsible for the radio halo emission must have energies $E_e \gtrsim$ a few GeV to radiate at frequencies $\nu \gtrsim 30$ MHz in order to reproduce the main properties of the observed radio halos (see, e.g., Colafrancesco & Mele 2001, and references therein). A few nearby clusters also show the presence of an EUV/soft X-ray excess (Lieu et al. 1999, Kaastra et al. 1999, 2002) and of an hard X-ray excess (Fusco-Femiano et al. 1999-2000; Rephaeli et al. 1999; Kaastra et al. 1999) over the thermal bremsstrahlung radiation. It is not yet clear whether these emission excesses may be produced either by Inverse Compton Scattering (hereafter ICS) of CMB photons off the relativistic electron population or by a combination of thermal (reproducing the EUV excess, Lieu et al. 2000) and supra-thermal (reproducing the hard X-ray excess by non-thermal bremsstrahlung, (e.g., Sarazin & Kempner 2000) populations of distinct origins. It is interesting in this context that in many of the clusters in which SZE has been detected there is also evidence for radio halo sources. So it is of interest to assess whether the detected SZE are in facts mainly from the thermal electron population or there is a relevant contribution from other non-thermal electron populations.

2.1 The non-thermal SZ effect in galaxy clusters

We derived elsewhere (Colafrancesco et al. 2003) a generalized expression for the SZE which is valid - in the Thomson limit - for a generic electron population in the full relativistic approach, includes the effects of multiple scattering and describes the SZE due to a combination of different electronic populations. Here we report only the main results of the derivation and we refer the interesting reader to a more detailed discussion given in Colafrancesco et al. (2003). The relevance of studying the non-thermal SZE is many-fold. First, it is unavoidably required in the study of the overall SZE for clusters showing non-thermal phenomena like Coma or A2163. Moreover, its possible detection can set constraints to the additional pressure contribution of the cluster atmosphere which is crucial to understand both the stability and the dynamics of the whole cluster. In turn, a detailed study of the non-thermal SZE in clusters can constrain the spectrum of the relativistic electron population. Such constraints are, in principle, more relevant than those inferred from the observation of separate non-thermal phenomena (like radio halos, hard X-ray and/or EUV excesses) because the non-thermal SZE is sensitive to the total pressure in the relativistic component which is a measure of the total spectrum of the high-E electrons and not of just a partial energy slice. Finally, the combined study of the non-thermal SZE with other non-thermal phenomena can efficiently constrain the origin of non-thermal phenomena in galaxy clusters.

Following the generalized derivation of Colafrancesco et al. (2003), the intensity change of the CMB due to the SZE can be written as

$$\Delta I(x) = 2\frac{(kT_0)^3}{(hc)^2}\, y\,\tilde{g}(x)\,, \tag{4}$$

where the Compton parameter is given by

$$y = \frac{\sigma_T}{m_e c^2}\int d\ell\, P\,. \tag{5}$$

Such a general expression takes into account both the case of a thermal electron population (the IC gas) with $P_{th} = n_e kT_e$ and the case of a non-thermal, relativistic electron population with $P_{rel} = n_{e,rel}\int_0^\infty dp\; f_e(p)\frac{1}{3}pv(p)m_e c$, where $f_e(p)$ is the momentum distribution of the considered electron population. The general form of the spectral shape $\tilde{g}(x)$ of the SZE has also the advantage to recover again both the thermal and the non-thermal cases as

$$\tilde{g}(x) = \frac{m_e c^2}{\langle kT_e\rangle}\frac{1}{\tau}\left[\int_{-\infty}^{+\infty} dsP(s)(i_0 x e^{-s}) - i_0(x)\right], \tag{6}$$

which is given in terms of the total redistribution function $P(s)$ (here $s \equiv ln\nu'/\nu$), the optical depth of the electron population τ and the undistorted CMB intensity $i_0(x)$ (see Colafrancesco et al. 2003 for technical details). The quantity $\langle kT_e\rangle = (\sigma_T/\tau)\int d\ell P$ is the generalization of the specific energy per particle kT_e of the thermal electron distribution.

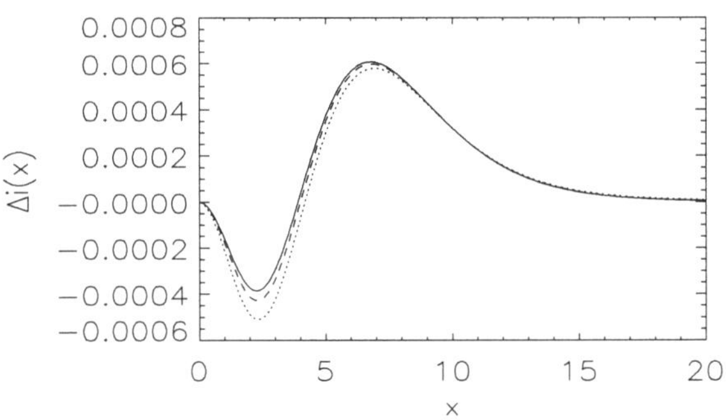

Figure 3. The spectral distortion, in units of $2(k_B T_0)^3/(hc)^2$, of a single thermal population (solid) is compared with that produced by a combination of thermal and non-thermal population with $p_1 = 0.5$ and $n_e(\tilde{p}_1 = 100) = 10^{-5}$ (dashes) and $3\cdot 10^{-5}$ cm^{-3} (dotted).

The spectra of the thermal and non-thermal SZE are distinctly different (see Figs.1-2) and the overall spectrum of the SZE measures the energy densities in

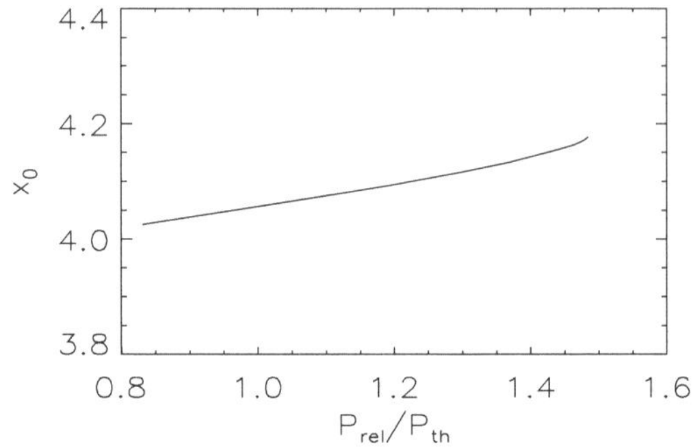

Figure 4. The behaviour of the zero of the total SZE produced by a combination of a thermal and non-thermal population with $n_{e,rel}(\tilde{p}_1 = 100) = 3 \cdot 10^{-5}$ cm^{-3} as a function of the pressure ratio P_{rel}/P_{th}.

the thermal and in the non-thermal electron populations, separately. The main spectral differences for the non-thermal SZE with respect to the thermal SZE are a maximum of the spectral shape which is moved to higher and higher frequencies for increasing values of the lower cutoff in the momentum distribution, p_1. As a consequence, also the null of the non-thermal SZE, x_0, is moved to higher frequencies for increasing values of p_1. A relevant result of our analysis shows that the location of the zero of the total SZE depends on the pressure ratio $\bar{P} = P_{rel}/P_{th}$ between the relativistic and thermal pressures of the two different electron populations. It increases non-linearly with $\bar{P}$ up to values $x_0 \sim 4.2$ for $\bar{P}$ up to values ≈ 1.5 (see Fig.4). This is a unique and relevant feature of the non-thermal SZE in clusters since it yields, in principle, a direct measure of the total pressure in relativistic non-thermal particles in the cluster atmosphere, an information which is not easily accessible from the study of other non-thermal phenomena like radio halos and/or EUV or hard X-ray emission excesses. Thus, the detailed observations of the frequency shift of x_0 provides unambiguously a constraint to the relativistic particle content in the IC medium. Also, such a measurement is crucial to determine the true amount of kinematic SZE in clusters since it is usually estimated from the residual SZ signal at the location of the zero of the thermal (relativistic) SZE.

We have also generalized our derivation of the total SZE in galaxy clusters to the case of a combination of different thermal electron populations (see Colafrancesco et al. 2003). Any additional cool IC gas component produces a tightening of the photon redistribution function, an increase in the total optical depth and hence a substantial change in the spectral distortion at the minimum and at the maximum of the SZE. The location of the zero of the SZE also decreases in frequency due to the presence of the cooler component which decreases the pressure ratio P_2/P_1. Thus, the possible detection of an additional cold component in the cluster atmosphere through observations of the total (thermal plus thermal) SZE, allow to test the possible thermal origin of the EUV excess observed in several nearby clusters.

3. The case of A2163

The non-thermal SZE may already have been observed as a part of the SZ signal from clusters which show non thermal phenomena like radio halos. Here we discuss the specific case of A2163 deferring the discussion of the Coma cluster to a separate paper (Colafrancesco, these Proceedings).
A2163 ($z = 0.203$) possesses a giant radio halo (Feretti et al. 2001), with diameter $\sim (2.9 \pm 0.1)h_{50}^{-1}$ Mpc, centered on the X-ray emission. The slope of the radio halo spectrum is $\alpha_r \approx 1.6 \pm 0.3$ corresponding to an electron spectrum $\alpha \approx 4.2 \pm 0.6$. It has been estimated from radio data at 1.365 and 1.465 GHz and is consistent with the results obtained by Herbig & Birkinshaw (1994) in the range 10 MHz – 10 GHz. There is no evidence of hard X-ray excess due to a non-thermal component in the BeppoSAX data of this cluster.
A strong SZE has been observed in A2163 from BIMA at 28.5 GHz (LaRoque et al. 2002), from DIABOLO at 140 GHz (Desert et al. 1998) and from SuZIE at 140, 218 and 270 GHz (Holzapfel et al. 1997; these data are dust-corrected in LaRoque et al. 2002) thus bracketing the null of the thermal SZE.
Here, we re-analyzed the data on A2163 trying to put constrains on the possible presence of a non-thermal SZE by fitting the available data with a combination of a thermal and non-thermal SZE.
The thermal population in A2163 has a temperature of $k_B T_e = 12.4 \pm 0.5$ keV and a central density of $n_{e,th} \simeq 6.82 \cdot 10^{-3}$ cm^{-3}. The parameters describing the spatial distribution of the IC gas are $r_c = 0.36\, h_{50}^{-1}$ Mpc and $\beta = 0.66$. With these values the optical depth towards the cluster center is $\tau_{0,th} = 1.56 \cdot 10^{-2}$ and the central Comptonization parameter is $y_{0,th} = 3.80 \cdot 10^{-4}$. The thermal SZE fits the data with a $\chi^2 = 1.71$ and hence it is statistically acceptable. However, the inclusion of a non-thermal component of the SZE is able to improve sensitively the fit to the data. We consider a non-thermal population with a double power-law spectrum with slopes $\alpha_1 = 0.5$, $\alpha_2 = 2.5$, $p_{cr} = 400, p_2 \rightarrow \infty$ and we set p_1 and the density $n_{e,rel}(\tilde{p}_1)$ as a free parameters. We also assume that the spatial distribution of the non-thermal population is similar to that of the thermal population as indicated by the extension of the radio halo in A2163.
A model in which $p_1 = 100$ and $n_{e,rel}(\tilde{p}_1) \approx 2.5 \cdot 10^{-5}$ cm^{-3} best fits the data with a $\chi^2_{min} = 1.05$, much lower than the $\chi^2_{min} = 1.71$ yielded by the single thermal population. The previous parameters point to a non-thermal population which carries a non negligible pressure contribution $P_{rel} \approx 0.29 P_{th}$ and which

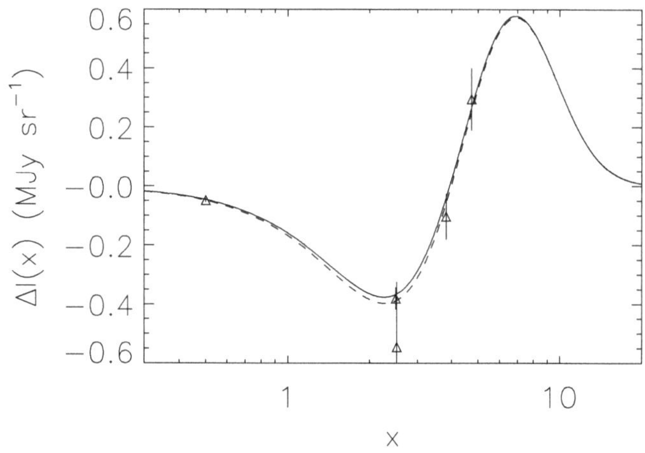

Figure 5. Theoretical expectations for the spectrum of the SZE in A2163. We show the fit to the available data yielded by a thermal population (solid curve) and the expectations obtained from a combination of thermal and non-thermal populations with $p_1 = 100$ for a value of the pressure ratio $P_{rel}/P_{th} = 0.29$ (dashed curve), which provides the best fit.

has a spectrum not extended at momenta lower than $p_{min} \simeq 100$, corresponding to $E_{min} \approx 50$ MeV. The SZ data favour also flat enough ($f_e \sim p^{-0.5}$) momentum spectrum below $p \sim 400$ to avoid destructive feedback effects on the thermal IC gas. Relativistic electrons with such flat energy spectrum do not produce a relevant extra heating and/or extra X-ray emission with respect to the IC gas in A2163. Also the IC energy losses of such relativistic electrons do not yield a substantial hard X-ray emission, in agreement with the available limit on A2163 obtained from BeppoSAX observations.

Our analysis shows that: *i)* a non-thermal SZE is produced by relativistic electrons producing radio halo emission has to be present in A2163. The χ^2 analysis indicates that its amplitude could be appreciable (see Fig.5) and corresponds to a pressure in relativistic particles $P_{rel} \approx 0.3 P_{th}$; *ii)* the possibility of having SZ observations with better precision can offer the possibility to disentangle between the thermal and any non-thermal component of the SZE; *iii)* the detection of a non-thermal SZE can set strong constrains on the nature of the non-thermal population and on its feedback on the thermal one. In these respects, it is appealing that the physical characteristics of the non-thermal population can be constrained through a detailed study of the SZE observed in the same galaxy cluster.

This example shows one of the potential uses of the SZE to obtain information

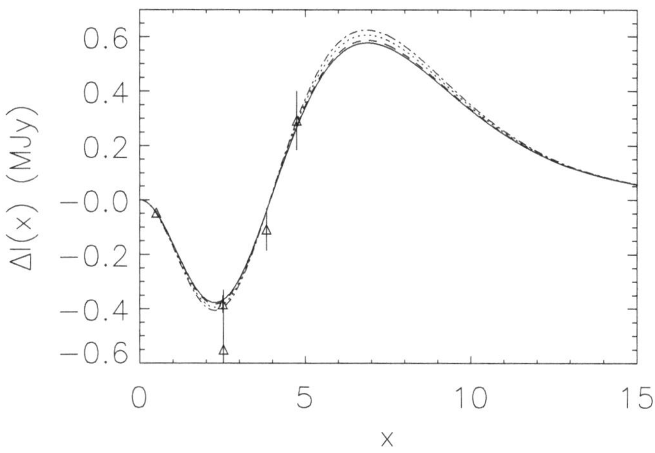

Figure 6. The spectrum of the SZE in A2163 obtained from a combination of two thermal populations with $k_BT_{e,1} = 12.4$ keV and $k_BT_{e,2} = 0.5$ keV. We show the cases of P_2/P_1: 0 (solid line), $5.91 \cdot 10^{-3}$ (dashed line), $2.09 \cdot 10^{-2}$ (dotted line, which yields the best fit), $2.96 \cdot 10^{-2}$ (dot-dashed line).

on the properties of different electronic populations which are residing in the atmospheres of galaxy clusters.

We also considered the calculation of the total SZE produced by a combination of two thermal electron populations, with the warm population temperature in the range $k_BT_{e,2} \sim 0.1 - 1$ keV and density in the range $n_{e,2} \sim 10^{-2}–10^{-4}$ cm^{-3}. We also considered two different spatial distributions of the hot and warm populations: in the first case the two populations have the same spatial extent and in the second one the cooler population is spatially more extended than the hotter one. In this last case we use parameters $r_{c,2} = 1.5\, r_{c,1}$ and $\beta_2 = 0.5$. The available SZ observations allow also in this case to set constrains on the parameters $n_{e,2}$ and $T_{e,2}$. A best fit which minimizes the χ^2_{min} is obtained for values $k_BT_{e,2} = 0.5 - 1$ keV (see Fig.6), while the case $k_BT_{e,2} = 0.1$ keV seems to be disfavoured by the data. Finally we want to remark that, in the set of models we considered in this section, the best fit to the data (with a $\chi^2_{min} \approx 1.05$) is obtained from a combination of thermal and non-thermal populations, a case which seems to be favoured by the present data.

Future SZ observations of A2163 with higher spectral resolution and sensitivity will allow to set stronger constraints on the non-thermal population. In particular, the possibility to measure accurately the frequency location of x_0 could yield direct information on the value of the pressure in relativistic particles, P_{rel}, confined in the cluster atmosphere and on their energy distribution.

4. Summary

Beyond the relevance of the study of the non-thermal SZE as a bias for the cosmologically relevant thermal SZE, the non-thermal SZE has also a crucial astrophysical relevance as a test for the presence of any additional population of electrons with both a non-thermal or a thermal energy spectrum. In fact, the non-thermal SZE actually measures the total pressure of the non-thermal electron population and hence yield constrains to its energy spectrum. Analogously, the additional thermal SZE produced by a cooler thermal component provides information on its temperature and density.

The specific spectral and spatial features of the non-thermal SZE can be derived through a multifrequency observation with high sensitivity and narrow-band detectors. The optimal observational strategy is to observe galaxy clusters in the frequency range $x \sim 2-8$ where the peculiar spectral features allow clearly to disentangle the non-thermal SZE from the thermal one. The PLANCK surveyor experiment has the capabilities to detect and map the non-thermal SZE in a large number of nearby radio-halo clusters. However, dedicated experiment with high sensitivity and narrow band spectral coverage are also adequate to detect the non-thermal SZE in radio-halo galaxy clusters.

References

Birkinshaw, M. 199, Phys.Rep., 310, 97

Carlstrom, J.E., Holder, G.P. & Reese, E.D. 2002, ARAA, 40, 643

Colafrancesco, S. & Mele, B. 2001, ApJ, 562, 24

Colafrancesco, S., Marchegiani, P. & Palladino, E. 2003, A&A, 397, 27

Desert, F. et al. 1998, New Astronomy, 3, 655

Feretti, L. et al. 2001, A&A, 373, 106

Fusco-Femiano, R. et al. 1999, ApJ, 513, L21

Fusco-Femiano, R. et al. 2000, ApJ, 534, L7

Herbig, T. and Birkinshaw, M. 1994, BAAS, 26, No. 4, 1403

Holzapfel, W.L. et al. 1997, ApJ, 481, 35

Kaastra, J. et al. 1999, ApJ, 519, L119

Kaastra, J. et al. 2002, ApJ, 574, L1

LaRoque S.J. et al. 2002, ApJ submitted (preprint astro-ph/0204134)

Lieu, R. et al. 1999, ApJ, 510, L25

Lieu, R. et al. 2000, A&A, 364, 497

Rephaeli, Y., Gruber, W. and Blanco, P. 1999, ApJ, 511, L21

Sarazin, C.L. and Kempner, J.C. 2000, ApJ, 533, 73

Sunyaev, R.A. and Zel'dovich, Ya.B. 1972, Comments Astrophys. Space Sci., 4, 173

Sunyaev, R.A. and Zel'dovich, Ya.B. 1980, ARA&A, 18, 537

WHAT THE SZ EFFECT CAN TELL US ABOUT THE ELECTRON POPULATIONS IN THE COMA CLUSTER

Sergio Colafrancesco
INAF- Osservatorio Astronomico di Roma
Via Frascati 33, I-00040 Monteporzio (Roma), Italy
cola@mporzio.astro.it

Abstract We discuss here the physical constraints on the thermal (warm and hot) and non-thermal electron populations in the Coma cluster that can be set from the most recent SZE observations.

Keywords: Cosmology, Galaxy clusters: Coma, CMB, Cosmic-ray interaction

1. The Coma cluster

The thermal Sunyaev-Zel'dovich effect (hereafter SZE, Sunyaev & Zel'dovich 1972, 1980) is a probe of free electron density structures in the universe at all redshifts from $z = 0$ to the recombination epoch provided that the electron overdensities and temperatures are sufficient to produce a detectable effect. In general, only galaxy clusters have dense enough atmospheres and high enough temperatures to provide measurable effects. Thus the SZE is a measure of the properties of galaxy clusters which hold substantial atmospheres. As such, the thermal SZE is a powerful probe of cosmology and of cluster evolution (see, e.g., Carlstrom et al. 2002 for a review).
However, many galaxy clusters show also evidence of non-thermal phenomena (see Colafrancesco, these Proceedings). The Coma cluster is widely considered as the archetype of a system consisting of such a complex electronic distribution for which the largest database at different wavelengths has been accumulated so far. A thermal bath of electrons at a temperature $k_B T_e \approx 8.2$ keV is responsible for the bulk of the X-ray emission in the energy range $1-10$ keV. Beyond the thermal X-ray emission, there is definite evidence of diffuse radio emission at frequencies $\nu_r \sim 30MHz - 5GHz$ (Thierbach et al. 2002). There is also evidence of an emission excess in the Extreme UV (EUV) (Lieu et al.

R. Lieu and J. Mittaz (eds.), Soft X-Ray Emission from Clusters of Galaxies and Related Phenomena, 147–152.

1996) and in the soft X-ray (Lieu et al 1999) energy bands. Finally, an emission in excess over the thermal bremsstrahlung has been also detected in the hard X-ray (HXR) energy range between 30 and 80 keV with the BeppoSAX-PDS instrument (Fusco-Femiano et al. 1999) towards the direction of Coma. At present only a 2σ upper limit of $F_{\gamma}^{EGRET}(> 100\ MeV) \approx 4 \cdot 10^{-8}\ cm^{-2}s^{-1}$ has been found at γ-ray energies from the EGRET instrument on board the CGRO satellite (Sreekumar et al. 1996).
While we have a clear understanding of the properties of the thermal electron population which is the main responsible of the thermal bremsstrahlung emission at X-ray energies $\sim 1 - 10$ keV, there is not yet a clear understanding of the origin (either non-thermal or thermal) of the additional electron populations.
The shape of the relativistic electron spectrum is set with good accuracy by the radio-halo data. The Coma radio halo spectrum in the range $\sim$ 30 MHz - 1.4 GHz requires electrons with energy in the range $\sim (2.8 - 19.4)(B/\mu G)^{1/2}$ GeV with a spectrum given approximately by a power-law

$$n_{rel} = n_0 \left(\frac{E}{GeV}\right)^{-x}, \tag{1}$$

where $x \approx 3.5$ (the uncertainty on this slope is $\sim 5\%$). Note, however, that the radio halo data cannot set, by themselves, the normalization of the spectrum because the radio halo synchrotron flux depends strongly also on another parameter, namely the IC magnetic field B, as $F_\nu \propto n_0 B_\mu^{(x+1)/2} \nu^{-\alpha_r}$, which is valid for a power-law spectrum as in Eq.1 with $\alpha_r = -(x-1)/2 \approx 1.25$ in the aforementioned frequency range. An uncertainty of a factor 2 in the value of B reflects in a change of a factor ~ 5 in F_ν (see Fig.1).
A direct estimate of the normalization of the relativistic electron spectrum could come from the BeppoSAX-PDS data in the energy range 20 - 80 keV under the assumption that the HXR excess observed in Coma is due to ICS of the CMB photons off the relativistic electrons with energies $\approx 3.3 - 5.5$ GeV, i.e. a fraction of those required to reproduce the radio halo spectrum. Assuming a spectral slope $x \sim 3.5$, the HXR excess in Coma can be recovered with a value $n_0 \sim 1.5 \cdot 10^{-9}\,\mathrm{cm}^{-3}\,\mathrm{GeV}^{-1}$ (see Fig.1). (Note that this value should be actually considered as an upper limit for the normalization n_0 of the relativistic electron spectrum since the nature of the HXR excess of Coma it is not yet well established. In fact, the $20 - 80$ keV flux observed with the PDS instrument might be contaminated by other hard X-ray sources not belonging to the cluster, like heavily obscured AGNs in the PDS field of view (Nevalaineen et al. 2003).) The extrapolation of the ICS spectrum which fits the HXR excess of Coma down to lower energies $\lesssim 0.25$ keV does not fit the EUV excess measured in Coma because it is too steep and yields too high flux compared to the measured flux by the EUV satellite in the $0.065 - 0.245$ keV band (see Fig.3). The EUV excess is more consistent with a

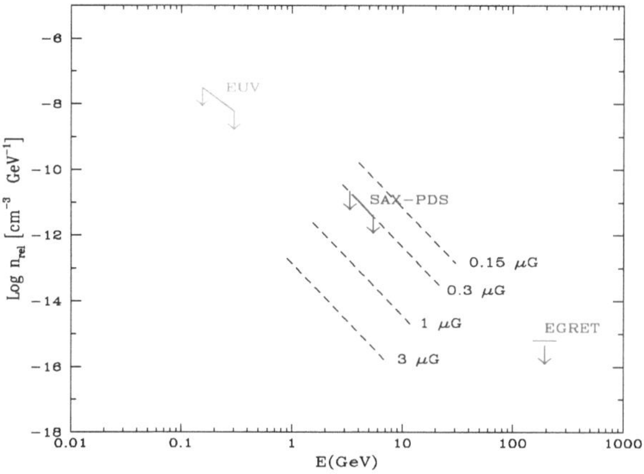

Figure 1. The constraints on the spectrum of relativistic electrons in Coma as obtained from observation of the radio halo spectrum (blue dashed lines), HXR excess (red solid line and arrows), EUV excess (green solid line and arrows) and γ-ray upper limit (magenta arrow).

power-law spectral slope $x \approx 2.5$ which is much flatter than the value $x \approx 3.5$ required by the radio halo and HXR spectra. Thus, under the assumption that the HXR of Coma is produced by ICS of CMB photons, the minimal requirement is that a break in the electron spectrum should be present in the range $0.3 - 2.8$ GeV in order to avoid an excessive EUV contribution by the relativistic electron ICS and to be consistent with both the radio halo spectrum and with the EUV excess of Coma. Another upper limit on the quantity $n_{rel}(E)$ is set at $E = 100$ MeV by the EGRET upper limit on the Coma γ-ray flux. Assuming the power-law spectrum given in Eq.(1) with $x \approx 3.5$, we found $n_0 \lesssim 5.7 \cdot 10^{-9}\ cm^{-3}\ GeV^{-1}$ (see Fig.1) obtained from the ICS γ-ray emission of the electrons with $E \gtrsim 190$ GeV in the high-energy tail of the spectrum. A more stringent limit obtains considering also the γ-ray emission produced by the relativistic bremsstrahlung of the electrons with $E \gtrsim 100$ MeV, given by $F_\gamma \approx 1.9 \cdot 10^{-11} B^{-2.3} (E/GeV)^{-3.6}\ pho\ cm^{-2}\ s^{-1}\ GeV^{-1}$ (see, e.g., Vestrand 1994).

2. The non-thermal SZE in the Coma cluster

In the context highlighted above, the SZE can be used as a powerful probe of the structure of the Coma cluster atmosphere since it is a sensitive probe of the overall pressure in each free electron species which is present along the line of sight through the cluster. The specific strategy to disentangle the thermal SZE from any other additional SZ signal of non-thermal and/or thermal origin is to have a wide spectral coverage of the SZ signal from the radio region to the sub-mm region thus bracketing the null of the overall SZE (see Colafrancesco

et al. 2003, Colafrancesco & Marchegiani 2003 for a detailed discussion of this point).
Recent radio and sub-mm data can probe now the SZ spectrum of the Coma cluster over a wide frequency range bracketing the null of the thermal SZE expected at $x \equiv h\nu/k_B T_0 \gtrsim 3.83$. The most recent data on the SZE in Coma are shown in Fig.2.

2.1 Contraints from the SZE data

In our re-analysis of the available SZ data on Coma, we consider various cases:
A) a combination of a thermal (with $k_B T \sim 8.2$ keV) and a non-thermal population;
B) a combination of two thermal populations: a hot electron population with $k_B T \sim 8.2$ keV and a cooler one having temperature in the range $k_B T_e \sim 0.1 - 1$ keV;
C) a combination of a thermal population with $k_B T \sim 8.2$ keV, a warm thermal population with $k_B T_e \sim 0.1 - 1$ keV and a non-thermal population with $p_1 \geq 1000$ which is responsible for the radio halo emission.
We first derive the fit to the available SZ data using a single thermal population with $k_B T = 8.2$ keV. Such a fit will determine the reference model for our analysis. We find that the best fit to the available SZ data in the case of a single thermal population yields $\tau_{th} = 4.9 \times 10^{-3}$ with a $\chi^2 = 1.68$. This result is fully consistent with the analysis of DePetris et al. (2002). We also consider in our analysis the effect of a radial peculiar velocity v_r of Coma which generates a kinematic SZE whose amplitude is given by $\delta T = -\tau \frac{v_r}{c}$, where the peculiar velocity has a negative (positive) sign for an approaching (receding) cluster. The actual limit on the peculiar velocity of Coma are $v_r = -72 \pm 189(\pm 378)$ km/s (Bernardi et al. 2002).
Case A). The fit to the available SZ data improves in the presence of a non-thermal electron population with a power-law spectrum with slope $x = 3.5$ only if $v_r < 0$ km/s; it reaches a minimum for $v_r - 450$ km/s. In this case the best fit value of the pressure of the relativistic electrons is $P_{rel} \sim (0.04 - 0.14)\ P_{th}$ for a lower momentum spectrum cutoff of $p_{min} \sim 10 - 10^3$. For these values of p_{min} the maximum relativistic electron density is $n_{rel} \lesssim 7 \cdot 10^{-3}\ n_{th}$, a value which can be consistent with the HXR BeppoSAX data only if $p_{min} \gtrsim 70$, or equivalently only for electron energies $E_{min} \gtrsim 35$ MeV. The combination of the SZ data and the HXR BeppoSAX data can exclude an electron spectrum with a single power-law $x = 3.5$ extended down to ~ 35 MeV.
Case B). Also here the presence of a negative peculiar velocity allows the fit to improve with the addition of a warm thermal component. The best fit obtains with $k_B T_{e,2} \approx 0.1$ keV and $n_{e,2} \approx (0.8 - 1.7)\ n_{th}$ for $v_r \lesssim -261$ km/s.

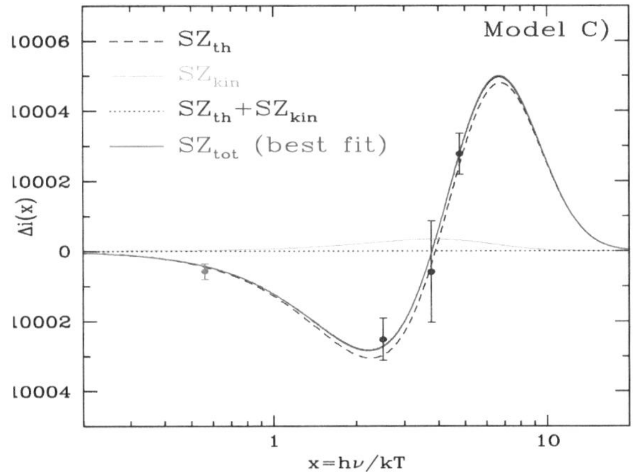

Figure 2. The best fit to the SZ data in the case C) discussed in the text is shown here (red solid curve) together with the prediction from a pure thermal electron population with $k_B T_2 = 8.2$ keV (blue dashed curve), the combination of the thermal and kinematic SZ effect (blue dotted curve) and the contribution from the kinematic component of the SZ effect evaluated for $v_r = -450$ km/s (green dotted curve). Data are from Herbig et al. (1995) and from DePetris et al. (2002).

Case C). The more realistic case of the presence of a relativistic electron population (with a spectrum consistent with the limits of Fig.1) and of an additional warm component yields a best fit to the SZ data for $n_{rel} \approx 10^{-6}$, $p_{min} = 10^3$ and $k_B T_{e,2} \approx 0.1$ keV with $n_{e,2} \approx 0.37\, n_{e,1}$. A 1 σ upper limit to the pressure of the warm component is given by $P_2 \lesssim 0.02 P_1$ which corresponds to an upper limit to the warm gas density of $n_{e,2} \lesssim 1.7\, n_{e,1}$. This results requires a peculiar velocity $v_r \approx -261$ km/s. Positive values of v_r worse the fit to the SZ data while more negative v_r values do not sensitively improve the fit.

3. Conclusions

A proper analysis of the SZ effect in galaxy clusters might provide a wealth of physical and cosmological information. Along this line, we have shown here that the recent SZ data on Coma provide useful constraints to the presence and to the spectrum of both relativistic electrons and warm thermal IC gas. We found that relativistic electrons in Coma cannot contribute with a substantial pressure $P_{rel} \lesssim$ a few % of the thermal pressure P_{th} in order to be consistent with the SZ data as well as with the EUV, HXR and gamma-ray limits. The SZ analysis gives a useful integral constraint to the overall energetics of the non-thermal electron population which is not accessible by other observational approaches. On the other hand, we have also shown that the SZ data favour the presence of an additional warm thermal component with temperature $k_B T \sim 0.1$ keV

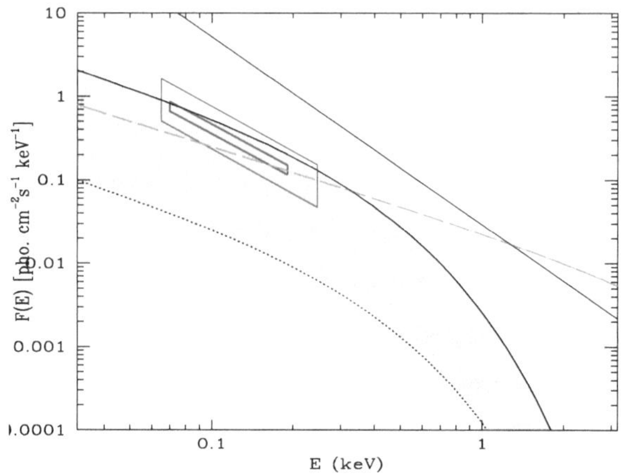

Figure 3. The spectrum of thermal electrons in Coma as constrained from SZE and EUV observations. The red thin area shows the EUV excess uncertainty region as given in Ensslin, Lieu and Biermann (1999) using a single slope $\alpha_{EUV} = 1.75$ while the red thick area shows the EUV excess based on the uncertainties given in Lieu et al. (1996) using again $\alpha_{EUV} = 1.75$. The green dashed curve is the thermal bremsstrahlung emission from the hot $T = 8.2$ keV electron population of Coma. The blue solid curve is the exrapolation at low energies of the ICS curve fitting the HXR excess in Coma. The yellow area encompassed by the black solid curves show the limits on the warm thermal component with primordial abundance set by our SZE analysis. Note that the upper thick solid curve correspond to the upper limit on the density of the warm gas allowed by the SZE data.

and density $n_e \sim (3.4 - 5.8) \cdot 10^{-3}\, h_{70}^{1/2} \mathrm{cm}^{-3}$ which superposes to the hot thermal component at the temperature $k_B T \approx 8.2$ keV. The warm component with almost primordial abundances could explain a substantial fraction of the EUV excess observed in Coma.

References

Bernardi, M. et al. 2002, AJ, 123, 2990
Birkinshaw, M. 199, Phys.Rep., 310, 97
Carlstrom, J.E., Holder, G.P. & Reese, E.D. 2002, ARAA, 40, 643
Colafrancesco, S., Marchegiani, P. & Palladino, E. 2003, A&A, 397, 27
De Petris, M. et al., 2002, ApJ, 574, L119
Fusco-Femiano, R. et al. 1999, ApJ, 513, L21
Herbig, T., Lawrence, C.R., & Readhead, A.C.S., 1995, ApJ, 449, L5
Lieu, R., Mittaz, J.P.D., Bowyer, S. et al., 1996, Science, 274, 1335
Neevalainen, J. et al. 2003, ApJ, submitted
Sreekumar, S. et al. 1996, ApJ, 464, 628
Sunyaev, R.A. and Zel'dovich, Ya.B. 1972, Comments Astrophys. Space Sci., 4, 173
Sunyaev, R.A. and Zel'dovich, Ya.B. 1980, ARA&A, 18, 537
Thierbach, M., Klein, U. and Wielebinski, R. 2003, A&A, 397, 53
Vestrand, T. 1994, see citation in Sreekumar et al. (1996)

V

THEORETICAL MODELS OF CLUSTERS, THE CSE AND THE WHIM

OBSERVATIONAL CONSTRAINTS ON MODELS FOR THE CLUSTER SOFT EXCESS EMISSION

Richard Lieu & Jonathan P.D. Mittaz
Department of Physics, UAH, Huntsville, AL35899

Abstract Although the cluster soft excess phenomenon is confirmed by XMM-Newton observations of many clusters, the cause of this new component of radiation remains an enigma. The two mechanisms proposed in the late 90's, viz. thermal emission from a massive warm baryonic gas and inverse-Compton scattering between cosmic rays and the microwave background, are still to date the only viable interpretations of the soft excess. In as much as cosmic rays cannot exist at the vast void of a cluster's outskirts, warm gas also cannot be present with any significant degree of abundance at the center of a cluster. In this sense, one could say that both models have their merits, and account for the soft excess in different spatial regions. There is however no clincher evidence that points definitively to the correctness of either explanation. Thus the door remains open for more exotic scenarios that could even consider the detected emission as signature of some dark matter process. In fact, the absence of absorption lines in the spectrum of background quasars along sightlines going through the outer radii of clusters argues against the thermal model within the domain where it most suitably applies. On the other hand, if the central excesses are due to cosmic rays, the pressure of the proton component will be large enough to choke a cooling flow, and missions like GLAST may have the sensitivity to detect the gamma rays that ensue from proton-gas interactions. However the diagnosis may turn out to be, it is likely that the new radiation represents something of cosmological importance.

1. Introduction

Historically, the cluster soft excess phenomenon was discovered by the EUVE mission from three clusters of galaxies: Virgo, Coma, and Abell 1795, when the EUV flux was detected at a level higher than that expected from the low energy spectral tail of the hot virialized intracluster medium (ICM) emission (Lieu et al 1996a,b; Mittaz, Lieu, & Lockman 1998). In the original interpretation of this excess, a warm intracluster gas component at $T \leq 10^6$ K was invoked, with a density radial profile at least as extended as, and a total mass comparable to, that of the hot ICM.

R. Lieu and J. Mittaz (eds.), Soft X-Ray Emission from Clusters of Galaxies and Related Phenomena, 155–162.

Such a model was not received with equal warmth, principally because of the high radiative cooling rate of the new gas. In order for a warm component to co-exist with the hot ICM, the former has to be clumped into clouds, so that pressure equilibrium can be maintained between the two media, i.e.

$$P_{warm} = P_{hot} \text{ requires } n_w = (n_h/10^{-3} \text{ cm}^{-3}) \left(\frac{T_h}{T_w}\right) \tag{1}$$

where $n_h \leq 10^{-3}$ cm^{-3} for the central and intermediate radii of clusters. Since the temperature ratio $T_h/T_w > 10$, Eq. (1) implies $n_w > 10^{-2}$ cm^{-3} we have, for these radii, a cooling time from free-free emission of

$$\tau_w = 6 \times 10^8 \left(\frac{T}{10^6 \text{ K}}\right)^{\frac{1}{2}} \left(\frac{n_w}{10^{-2} \text{ cm}^{-3}}\right)^{-1} \text{ years.} \tag{2}$$

Hence the gas is highly unstable. The difficulty may somewhat be alleviated by invoking magnetic fields in a generalization of Eq. (1):

$$P_w + \frac{B_w^2}{8\pi} = P_h + \frac{B_h^2}{8\pi}, \tag{3}$$

though it is still quite hard to envision field arrangements that can make a great difference. Another hurdle that confronts this type of scenarios is that photo-ionization of the warm clouds by the hot ICM radiation, which occurs at the very short timescale of:

$$\tau_{photo} \approx 2 \times 10^7 (F/10^4 \text{ ph cm}^{-2} \text{ s}^{-1})^{-1} \text{ yr} \tag{4}$$

for O VII, where F is the hot ICM emitted flux.

All the above was 6-8 years ago. Today, with the advent of XMM-Newton the warm gas is now thought to exist at large radii, accounting for the extended soft excess radial profile - we shall return to this point. Historically, however, affairs twisted and turned like a mountain road - an altogether distinct emission mechanism was considered instead, which also turned out to have a high probability of being relevant to the soft excess. Several authors independently and contemporaneously proposed the non-thermal model (Hwang 1997, Ensslin & Biermann 1998, Sarazin & Lieu 1998), in which the soft X-ray and EUV photons are produced by inverse-Compton scattering between a population of relativistic electrons and the cosmic microwave background. The basic tenet of the theory is that electrons capable of doing this have a Lorentz factor

$$\gamma \sim 300\ (h\nu/75eV)^{\frac{1}{2}}, \tag{5}$$

and their main loss mechanism electrons is indeed through this same inverse-Compton process, which restricts the electron lifetime to:

$$\tau_{IC} \sim 7.7 \times 10^9 (\gamma/300)^{-1} \text{ years} \tag{6}$$

Thus even if the electrons belong to a relic population that dates back to the era of cluster formation when there were considerably more AGN activity and shock acceleration, some of them survive to the present time, and can in principle account for the observed emission. In reality, however, the brightness level of the soft excess necessitates a large energy density of cosmic rays, which may continuously be produced in the centers of clusters.

2. Thermal origin of the soft excess - the missing baryons

From theoretical considerations (Cen & Ostriker 1999) one expects the majority of the baryons at the present epoch of the Universe's evolution to be 'hidden' in the form of a Warm-Hot-Intergalactic-Medium (WHIM) gas. Crudely speaking the argument goes as follows. If λ is the characteristic wavelength of the bulk motion of the intergalactic medium at a given epoch, then when these waves collide and break the thermal velocity of the shocked gas will typically be $v \sim H_o\lambda$ where H_o is the Hubble constant, i.e.

$$v \sim 100(H_o/70)(\lambda/1.5\ \mathrm{Mpc})\ \mathrm{km\ s^{-1}} \tag{7}$$

where we set λ at $\sim$ the present size (virial radius) of galaxy clusters, i.e. $\lambda \sim$ 1-3 Mpc. The value of v as given in Eq. (7) corresponds to a gas temperature of 10^{5-7} K.

The assumption of a WHIM origin alleviates several of the aforementioned problems with a thermal model, notably the pressure problems and the cooling time. But is there observational evidence both for the WHIM and its association with the cluster soft excess (CSE)?

2.1 Characteristics of the soft excess as thermal emission

If we assume that the CSE is indeed thermal and from the WHIM, what are the constraints, if any, that we can make on the parameters for the warm baryonic gas that resides in the vast domain of intergalactic space outside the virial radius of clusters? For example, it is deceptive to think that an arbitrarily low density can be envisaged without eventually contradicting certain observational facts.

To pursue this point further, we take as specific example the Coma cluster, which was seen by the ROSAT/PSPC to possess a giant soft emission halo extending to $\geq$ 1 Mpc (Bonemente et al. 2003). There is no plausible explanation of such a phenomenon other than the thermal model with a warm and massive baryonic component. The observed brightness of the soft excess may be expressed as the projected emission integral (EI) of the warm gas column, i.e.

$$\mathrm{EI} = n^2 LA, \tag{8}$$

where n is the HI number density, assumed to be uniform throughout an optically thin emission column of length L and cross sectional area A. For the 0-20 arcmin central radius of Coma, $EI \sim 10^{68}$ cm^{-3} from the published ROSAT data (Bonamente et al 2003), leading to:

$$L \approx 3(n/10^{-3}cm^{-3})^{-2}\ \mathrm{Mpc} \tag{9}$$

and an equivalent HI column density $N_H = nL$ of

$$N_H \approx 10^{22}(n/10^{-3}cm^{-3})^{-1}\ \mathrm{cm}^{-2} \tag{10}$$

A warm column with such a value of N_H already poses significant opacities to certain EUV and soft X-ray lines that may originate from underlying emission layers, though its continuum opacity remain small. *If the density n assumes values smaller than* $\sim 10^{-3}cm^{-3}$, the resulting column will be too high for consistency with observations of bright QSOs behind Coma (see Figure 1).

The second constraint comes from the total mass budget of the warm gas. To make the point, let us consider the simple picture of a warm halo as having the uniform density n and radius L as above. Then the total mass M is given by

$$M = \frac{4}{3}\pi L^3 n m_p f \tag{11}$$

where m_p is the proton mass and f is the volume filling factor of the gas, i.e. fraction of the spherical volume occupied by the warm filaments, which might be in the form of 'spaghetti' converging onto a node - the cluster. If there exists an inner radius r_o at which the filamentary footpoints cover the entire cluster surface, and if the filaments have constant cross sectional area, then, by geometry, $f = (r_o/L)^2$, and

$$M = 10^{14}(n/10^{-3}cm^{-3})^{-1}(r_o/0.5\ \mathrm{Mpc})^2\ M_\odot \tag{12}$$

where in arriving at Eq. (12) use was made of the observed soft excess emission brightness, Eq. (9). Once again, it can be seen that if the density is too low, or the warm gas located too far out (r_o large to avoid contact with the hot ICM), the resulting mass budget will exceed even that of the dark matter limits placed by gravitational lensing and galaxy velocity dispersion measurements.

The final constraint comes from the cooling time of the gas. As shown by Eq. (2) if the density becomes too high then the gas will catastrophically cool. For Coma we have a gas at a temperature of 0.2 keV and an abundance of 0.1 (Bonamente et al 2003) setting a limit on the density of the gas $n < 6,5 \times 10^{-4}$ cm^{-3} for the gas to remain stable in a Hubble time. However, at these temperatures the cooling time is also strongly affected by line emission. Using the emissivities taken from Sutherland & Dopita (1993) we then derive

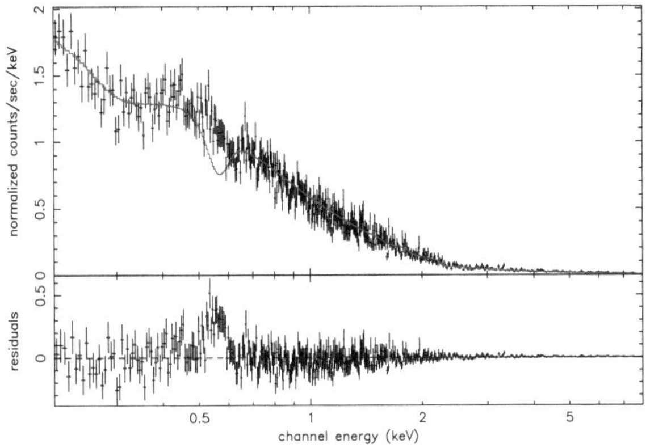

Figure 1. The EPIC PN spectrum of X-Comae together with the best fit model (a power-law plus Galactic absorption) and a gaussian absorption line with an equivalent width of 28eV. This line is rejected at the 99% confidence level by an F-test demonstrating that the strong absorption expected from a WHIM like model for the CSE is not seen.

a revised cooling time of

$$\tau_w = 1.18 \times 10^9 \left(\frac{n}{10^{-3} cm^3}\right)^{-1} years \qquad (13)$$

This sets a tighter limit on the density of $n < 8.64 \times 10^{-5}$cm^{-3} and hence from Eq. (10) a limit on the column density for the CSE of $> 10^{23}$cm^{-2}. Then using Eq. (8) of Nicastro et al. (1999) and the gas parameters for Coma we can derive a limit on the equivalent width of the OVII 21.6Å absorption line of $EW > 160$eV.

Such a large equivalent width should be observable. In Figure 1 is shown an EPIC PN spectrum of X-Comae, a quasar located 28 arcminutes from the Coma cluster center and at a redshift of 0.091, well behind the Coma cluster. The plot also shows the best fitting model (a power-law with Galactic absorption) together with an absorption line at the energy of OVII 21.6Å with an equivalent width of 28eV. As is apparent from the spectrum, a line with this equivalent width is not consistent with the data. This lack of absorption sets a limit on the observed column density of gas to a factor of at least 5 lower than the column needed to stabilize the gas for the duration of a Hubble time. If we reduce the cooling time to match up the two constraints we then find that the warm CSE emitting component could only have been formed 2.7Gyr ago, or at a redshift of 0.23. This is clearly in contradiction with models for the formation of the WHIM (e.g. Dave et al 2001).

2.2 Characteristics of the WHIM from UV absorption lines and its relationship to the CSE

The above discussion shows that, in fact, strong constraints can be placed on the thermal/WHIM interpretation of the CSE. It is then reasonable to ask what is the evidence of a WHIM at all? There are plenty of examples of absorption lines in the UV from the intergalactic medium in the literature, most notably those from the Ly-α forest. However, in general the lines being studied (Ly-α, Mg II, CIV for example) probe clouds of relatively low temperature when compared to those of the WHIM. With the launch of FUSE this has changed because FUSE is able to look for absorption from e.g. the OVI 1031.9, 1037.6Å doublet which is sensitive to material with a temperature typical of the WHIM ($10^5 - 10^7$K). Current observations show that there is indeed material in intergalactic space at a density and temperature consistent with the WHIM, although most detections are from material within our local group (e.g. Nicastro et al. 2003 and references). There are also several reported detections of OVI absorption at intermediate redshifts (Tripp & Savage 2000, Tripp et al. 2000). These observations allow us to conclude that material showing all the characteristics of the WHIM does exist.

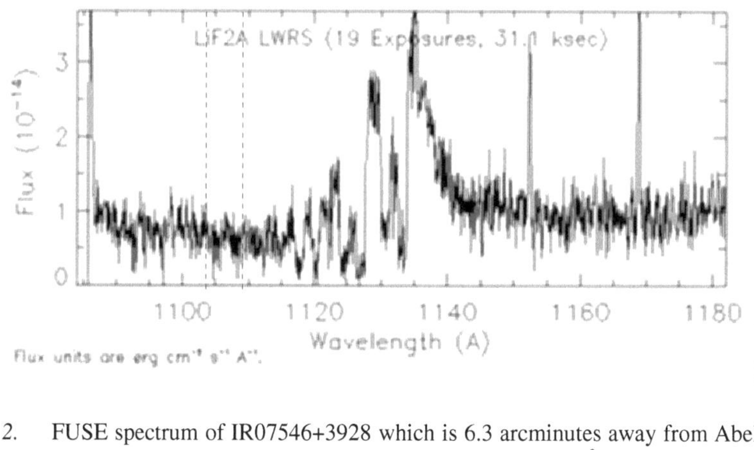

Figure 2. FUSE spectrum of IR07546+3928 which is 6.3 arcminutes away from Abell 607. The dashed lines show the expected position of the OVI1031.9, 1037.6Å absorption lines (spectrum taken from the data preview for the FUSE mission)

However, in terms of WHIM related to the CSE i.e. WHIM like material close to cluster of galaxies, the situation is less than clear. In general when absorption is seen at the redshift of a cluster the inferred column density of material is much lower than that required to explain the CSE. For example, a absorption system seen in the spectrum of 3C273 at a redshift of $z_{abs} = 0.00337$ (close to the redshift of the Virgo cluster, $z = 0.0036$) but located $>$ 3Mpc from the cluster center show material with a temperature $T > 10^{5.29}$K and has a column density of $log_{10}(OVI) \sim 13.32$ (Tripp et al. 2002). However, the typical OVI column density inferred from the emission properties of the CSE indicate that $log_{10}(OVI) = 16 - 18$ (e.g. Bonamente & van Dixon 2004). Other searches for absorption from nearby clusters have also met with difficulties. In another case a possible OVI absorption line seen at the redshift of Abell 3782 gives a column density associated with the cluster of $log_{10}(OVI) = 13.95$, which is again much smaller (factors of $10^2 - 10^4$) than the columns derived from the emission properties of the CSE (Bonamente & van Dixon 2004). Other AGNs behind clusters show no evidence for OVI absorption in their spectra (see for example Figure 2). It would thus appear that as in the the case of X-ray absorption, UV absorption line studies in the vicinity of clusters are failing to find sufficient material to account for the CSE. It should be noted, however, that the number of available observations of QSOs behind clusters is still very small, and the temperature range accessible by these UV observations ($10^5 - 10^6$K) is lower than that observed for the CSE (typically kT = 0.2 keV e.g. Kaastra et al. 2003).

3. Non-thermal processes

By using the emission characteristics of the observed CSE we have shown that stringent limits can be placed on both thermal and non-thermal models for the CSE. In the case of the former, cooling time and pressure balance arguments imply that the only viable option is to associate the CSE in a WHIM like filament external to the cluster environment. While such an undertaking may alleviate many problems, the allowed range of column density for the warm gas means strong absorption lines are expected to be visible in the spectra of background lighthouses, which have not been seen in either the UV or soft X-ray regime. This contradiction is a significant embarrassment for the thermal interpretation of the CSE.

Turning now to non-thermal considerations, it is looking increasingly certain that this may be the only option for the center of clusters where the extreme brightness of the CSE necessitates an unacceptably large quantity of warm gas if the phenomenon is to be interpreted thermally. Also, in terms of a χ^2 goodness-of-fit to the XMM spectra (which is controlled by the continuum shape as the O VII line is too weak) there is as yet no compelling reason to prefer a thermal model. In fact, often for these central cluster regions the best-fit power laws are found to secure lower χ^2 values.

At the time of writing, the authors found that if non-thermal processes are responsible for the central soft excesses of A1795 and AS1101, which are bright enough to shine above the peaked emission of the 'cooling flow' gas in the hot ICM, the cosmic ray pressure will assume non-trivial values. A scenario may therefore be envisaged whereby the protons, which carry the bulk of the pressure, are in equipartition with the hot ICM. While these protons have the prerequisites to choke a cooling flow, they also produce secondary electrons (via pions) during interactions with the hot ICM gas. It is these electrons which undergo inverse Compton scattering with the microwave background photons, resulting in excess EUV and soft X-ray emissions.

4. Conclusion and Future prospects

We conclude that while the new observations by the XMM/Newton mission have convincingly demonstrated the existence of the CSE, the emission mechanism remains an unanswered question. The current picture attributes the phenomenon to cosmic rays at the center and thermal warm gas at the outer parts, but there are difficulties.

In the future, it is anticipated that the Astro-E2 and Constellation-X missions will deliver clincher evidence, by means of their superior spectral resolution, for or against the thermal origin of the soft excess, as the issue over the reality of the O VII lines will be resolved after their launches. Moreover, if cluster cosmic rays are present in abundance, they will produce gamma rays with the hot ICM - radiation which the GLAST mission is sensitive to. Observations of clusters via the Sunyaev-Zeldovich effect may also lead to a spectral distinction between thermal and non-thermal processes, as both carry significant SZ implications (it is sometimes argued that warm gas columns may not contribute much to the SZ integral $\int nkTdl$ because of the lower T, yet as we saw above this could be compensated by the larger column length L).

References

Bonamente, M., Lieu, R & Joy, M., 2003, ApJ, 595, 722
Cen, R. & Ostriker, J., 1999, ApJ, 514, 1
Davé et al., 2001, ApJ, 552, 473
Ensslin, T.A, Lieu, R. & Biermann, P.L., 1999, A&A, 397, 409
Hwang, C, 1997, Science, 278, 1917
Lieu, R., et al., 1996a, ApJLett, 458, 5
Lieu, R., et al., 1996b, Science, 274, 1335
Mittaz, J.P.D, Lieu, R. & Lockman, J., 1998, ApJLett, 419, 17
Nicastro, F., 2003, astro-ph/0311162
Sarazin, C. & Lieu, R., 1998, ApJLett, 494, 177
Tripp, T. et al., 2000, ApJLett, 524, 1
Tripp, T. & Savage, B, 2000, ApJ, 542, 42
Tripp, T. et al., 2002, ApJ, 575, 697

HIGH-RESOLUTION SIMULATIONS OF CLUSTERS OF GALAXIES

Daisuke Nagai[1,2], Andrey V. Kravtsov[1,2]

[1]*Center for Cosmological Physics, University of Chicago, Chicago IL 60637*

[2]*Department of Astronomy and Astrophysics, University of Chicago, 5640 S Ellis Ave, Chicago IL 60637*

Abstract Recently, high-resolution *Chandra* observations revealed many complex processes operating in the intracluster medium such as cold fronts, cooling flows with puzzling properties, and subsonic turbulent motions. The detailed studies of these processes can provide new insights into the cluster gas properties and cluster evolution. Using very high-resolution Adaptive Mesh Refinement (AMR) gasdynamics simulations of clusters forming in the CDM universes, we investigate structural evolution of the intracluster gas and the role of dynamical (e.g., mergers) and physical (e.g., cooling, star formation, stellar feedback) processes operating during cluster evolution. In particular, we use the simulations to investigate the structure and dynamical properties of cold fronts. We also discuss the internal turbulent motions of cluster gas induced by mergers. Finally, we present preliminary results from an ongoing program to carry out a series of simulations that include cooling, star formation, stellar feedback, and metal enrichment and discuss implications for the cluster observables.

Keywords: cosmology: theory – intergalactic medium – methods: numerical – galaxies: clusters: general – instabilities–turbulence–X-rays: galaxies: clusters

1. Introduction

Cosmological N-body+gasdynamics simulations are a powerful tool for modeling cluster formation and studying processes that determine observable properties of intracluster medium (ICM). The simulations start from theoretically-motivated initial conditions and follow the dynamics of gravitationally dominant dark matter and gasdynamics of baryons. Such *ab initio* simulations can capture the full complexity of matter dynamics during the hierarchical build-up of structures and, therefore, allow us to systematically study the effects and the relative importance of various processes operating during cluster evolution in a realistic cosmological setting.

R. Lieu and J. Mittaz (eds.), Soft X-Ray Emission from Clusters of Galaxies and Related Phenomena, 163–170.

We present analysis of a suite of simulations of cluster formation using the newly developed Adaptive Refinement Tree (ART) N-body+gasdynamics code Kravtsov, 1999; Kravtsov et al., 2002 in the currently-favored flat ΛCDM cosmology ($\Omega_0 = 0.3$, $h = 0.7$, $\sigma_8 = 0.9$). The code uses a combination of particle-mesh and shock-capturing Eulerian methods for simulating the evolution of DM and gas, respectively. High dynamic range is achieved by applying adaptive mesh refinement to both gasdynamics and gravity calculations. The peak spatial resolution in the cores of clusters will reach $\sim 1 - 5h^{-1}$ kpc and clusters will have $\gtrsim 10^6$ particles within the virial radius. This is more than an order of magnitude improvement over previous studies. The simulations thus bridge the gap between the superb resolution of current X-ray observations and the spatial resolution of numerical simulations and, hence, enable extensive comparisons of numerical simulations with the high-resolution *Chandra* observations. Such comparisons will allow us to understand underlying processes responsible for observed properties of ICM and help facilitate the interpretation of observations. Below, we illustrate this using three specific examples. The details of the studies can be found in Nagai & Kravtsov (2003) and Nagai, Kravtsov & Kosowsky (2003).

2. Cold Fronts in CDM cluster

Recent discoveries of *"cold fronts"* by *Chandra* observations have come as a surprise Markevitch et al., 2000; Vikhlinin et al., 2001a; Sun and Murray, 2002; Forman et al., 2002; Markevitch et al., 2002. It was quickly realized that the existence of such sharp features puts interesting constraints on the small-scale properties of the intracluster medium (ICM), including the efficiency of energy transport Ettori and Fabian, 2000 and the magnetic field strength in the ICM Vikhlinin et al., 2001b. However, placing meaningful constraints on the physics of the ICM requires detailed understanding of gas dynamics in the vicinity of cold fronts, since the power of these constraints relies on the validity of the underlying assumptions of the dynamical model. It is therefore interesting to search for counterparts of the observed *"cold fronts"* in simulated clusters forming in hierarchical models and study their structure and dynamical properties.

Sharp features similar to the observed cold fronts are produced naturally in a cluster merger when the merging subclump, moving slightly supersonically, is undergoing tidal disruption. The Figure 1 shows the photon count *Chandra* map (top-left) and the emission-weighted temperature map (bottom-left) of the region around the sharp feature identified in a simulated cluster during a major merger. The cold front is a low temperature region embedded in the high temperature gas just behind the left bow shock. The appearance and the spatial extent of the cold front in simulation ($\sim 0.5h^{-1}$ Mpc) are very

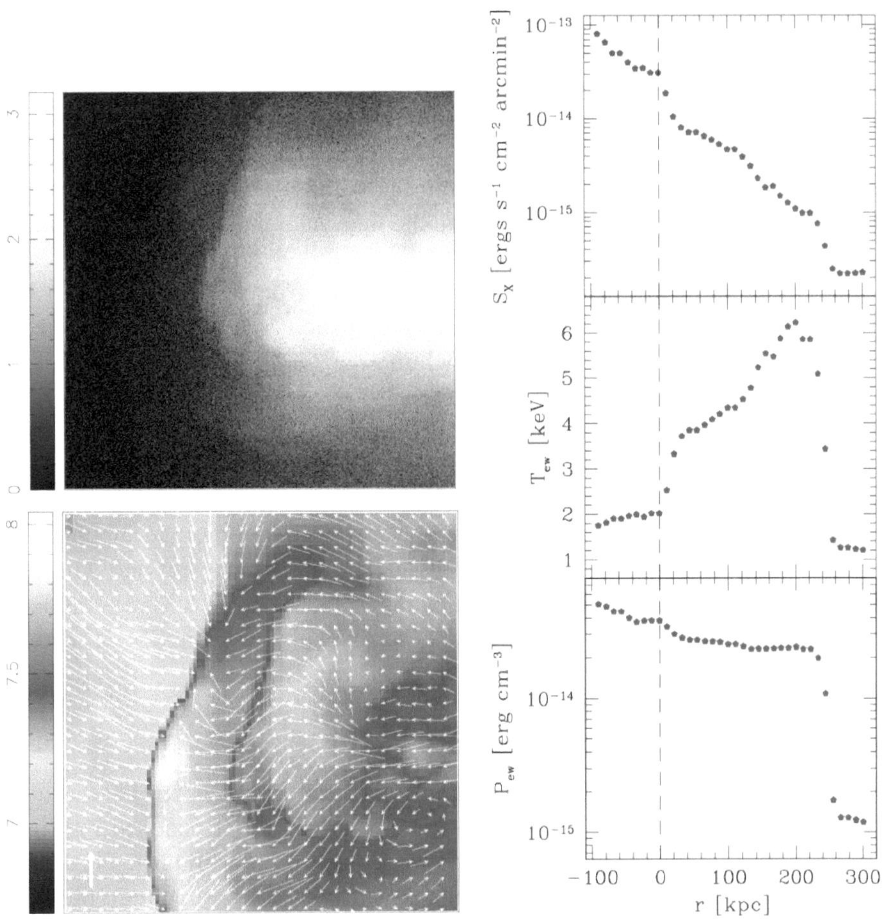

Figure 1. Cold front in the simulated ΛCDM cluster identified in a major merger at z=0.43. The top-left panel shows the mock $16' \times 16'$ ACIS-I *Chandra* photon count image of the cold front. The image was constructed assuming 50 ksec exposure in the $0.5-4$ keV band, cluster redshift of $z = 0.05$, and a flat background level of 4.3×10^{-6} cnts s^{-1} pixel^{-1}. The bottom-left panel shows the velocity map overlaid on the emission-weighted temperature map of the same region. The length of the thick vertical vector in the bottom left corner corresponds to 1000 km s^{-1}. The right panel shows the X-ray surface brightness, emission-weighted temperature, and emission-weight pressure profiles across cold fronts. The distance in kpc is measured relative to the cold fronts shown by dashed lines. The simulated profiles and images reproduce all the main features of the observed cold fronts (sizes, jumps in surface brightness and temperature, relatively smooth gradient of pressure across the front).

similar to those of two prominent observed cold fronts in clusters A2142 and A3667 Markevitch et al., 2000; Vikhlinin et al., 2001a. The right panel shows that the behavior of the profiles (opposite-sign jumps in surface brightness and temperature profiles and relatively smooth change of pressure across this front) reproduces those of observed cold fronts Markevitch et al., 2002. However, the velocity map around the simulated cold front (bottom-left) shows that the flow of gas is not laminar and in general is not parallel to the front. This is very different from assumptions made by Vikhlinin et al., 2001b, who assumed a laminar flow similar to that about a blunt body and used the sharpness of the observed front to put tight constraints on the thermal conduction and magnetic field strength in the vicinity of the front.

In the hierarchical models, "small-scale" cold fronts are expected to be much more common. Our simulations show that such cold fronts are produced in a minor merger when a merging subclump reaches the inner regions of the larger cluster and gas in front of the subclump is significantly compressed by ram-pressure (shown in the left panel of Figure 2). The compression sharpens and enhances the amplitude of gas density and temperature gradients across the front. The relatively cooler gas of the merging sub-clump is often trailed by low-entropy ($\sim$1-2 keV) intergalactic gas in the direction of the subclump's motion. Many instances of such "small-scale" cold fronts Sun and Murray, 2002; Forman et al., 2002; Markevitch et al., 2002 and diffuse X–ray filament extending beyond the outskirt of a cluster Durret et al., 2003 have been seen in *Chandra* and *XMM-Newton* observations of nearby clusters.

3. Internal turbulent motions of cluster gas

Frequent minor mergers discussed above will continuously stir the gas and generate turbulent gas motions within the cluster. The left panel of the Figure 2 shows the internal turbulent motions of cluster gas within the cluster in a relatively relaxed state. In the left panel, a small merging subclump moving with velocity $\gtrsim$ 1000 km s^{-1}along the filament is on the first approach to the cluster center and has not yet suffered major tidal disruption. When this merging subclump reaches the cluster core it generates random, slightly supersonic motions in which it dissipates its kinetic energy. The velocity is quite chaotic with typical motions at a level of $\sim 200 - 300$ km s^{-1}(typically 20-30% of the sound speed) even in the core of the relatively relaxed cluster due to frequent minor mergers (shown in the right panel). Our simulations suggest that regions of fast-moving gas are generally present even in "relaxed" clusters due to ongoing minor mergers. Gas motions of a similar magnitude in the cores of "relaxed" clusters are also implied by the *Chandra* observations Markevitch et al., 2002. This means that some processes (e.g., motion of substructures or asymmetric inflow of gas along filaments) in simulations continuously stir the gas even when the cluster is in hydrostatic equilibrium globally. Such gas motions have very important implications for the mixing and transport processes

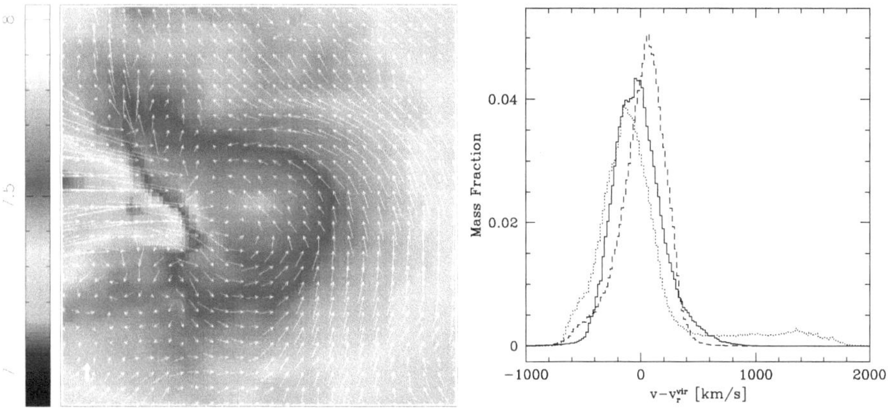

Figure 2. Internal gas flows of the cluster in a relatively relaxed state, undergoing a minor merger. Left: the velocity field is overlaid on the color maps of the emission-weighted temperature, color-coded on a log10 scale in units of Kelvin. The vertical vector in lower left corner correspond to 500 km/s. The size of the shown region is $0.78h^{-1}$ Mpc. Right: the distribution of the gas velocity component along three orthogonal projections within the virial radius of the cluster. The gas is moving with velocities of 200-500 km/s (typically 20-30% of the sound speed) even in the cores of relaxed clusters due to frequent minor mergers. Such gas motions have important implications for the mixing and transport processes in cluster cores, mass modeling of cluster, as well as observable properties of ICM. From Nagai, Kravtsov & Kosowsky (2003).

in cluster cores, mass modeling of cluster, as well as observable properties of ICM.

4. Effects of cooling and starformation

The Figure 3 shows the three-dimensional radial profiles of over-density of dark matter, total mass, entropy and temperature (clockwise from top-left) of the simulated clusters with adiabatic gasdynamics (dashed) and the simulations that include radiative cooling and starformation (solid). The cooled low-entropy gas is replaced by the higher entropy gas, which causes the increase of entropy and temperature in the center within 500kpc (roughly half of the virial radius). Significant fraction ($\sim 80\%$) of cooled gas within the core of the cluster has been turned into stars which formed a giant elliptical galaxy with mass of $M \gtrsim 10^{13} M_{\odot}$. Note also that the compression of baryons causes steepening of the dark matter density profile from $\sim r^{-1.5}$ to $\sim r^{-2.0}$ in the center of the cluster. Our simulations indicate that cooling and star formation are likely important processes that shape the properties of ICM as well as the dark matter profiles in the cluster core region.

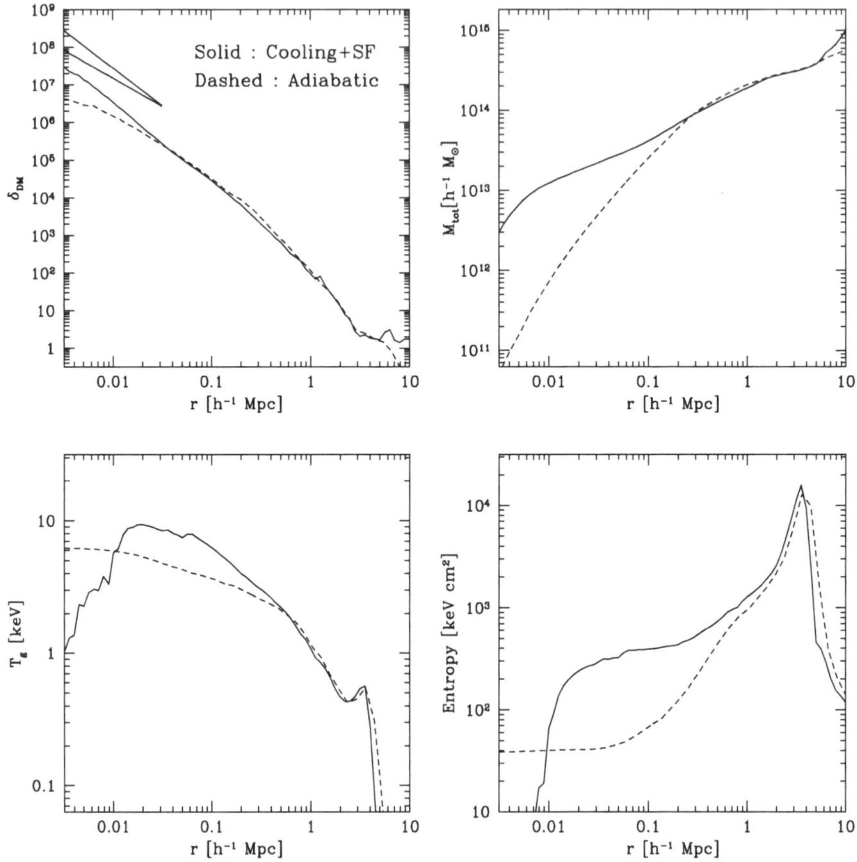

Figure 3. The 3D radial profiles of over-density of dark matter, total mass, entropy and temperature (clockwise from top-left) of the simulated clusters with adiabatic gasdynamics (dashed) and the simulations that include radiative cooling and starformation (solid). The cooled low entropy gas is replaced by the higher entropy gas, which causes the increase of entropy and temperature in the center within 500kpc (roughly half of the virial radius). These results support the suggestion of Voit and Bryan (2001) that gas cooling and star formation may be able to explain the observed properties of clusters. Note also that the cooling of baryon results in a steepening of the dark matter density profile in the center of the cluster. Two straight solid lines in the top-left panel indicate lines with $\sim r^{-1.5}$ and $\sim r^{-2.0}$.

5. Summary

We used the very high-resolution Adaptive Mesh Refinement (AMR) gasdynamics simulations of clusters forming in the CDM universe to investigate structural evolution of the intracluster gas and the role of dynamical (e.g., mergers) and physical (e.g., cooling, star formation, stellar feedback) processes operating during cluster evolution.

We find that the sharp features similar to the observed *cold fronts* are produced naturally in a cluster mergers. The simulated profiles and images reproduce all the main features of the observed cold fronts (sizes, jumps in surface brightness and temperature, relatively smooth gradient of pressure across the front). We find that the velocity fields of gas surrounding the cold front can be very irregular which would complicate analyses aiming to put constraints on the physical conditions of the intracluster medium in the vicinity of the front.

Although the most spectacular cold fronts are produced in a major merger, "small-scale" cold fronts arising from minor mergers are expected to be much more common, and many more instances of such cold fronts are expected to be found by *Chandra* observations of nearby clusters. The frequent minor mergers will also continuously stir the gas and generate subsonic turbulent gas motions within the cluster in a relatively relaxed state. Such gas motions, if present, have very important implications for the mixing and transport processes in cluster cores, mass modeling of cluster, as well as observable properties of ICM.

Finally, we presented preliminary results from an undergoing program to carry out a series of simulations that include cooling, star formation, stellar feedback, and metal enrichment. Using these simulations, we illustrated that cooling and star formation have a significant impact on the properties of ICM (e.g., entropy and temperature profiles) as well as dark matter distribution in the cluster core. Our results are consistent with the suggestion that gas cooling and star formation together increase both entropy and temperature in the cores of groups and clusters of galaxies (e.g., Voit and Bryan 2001). The results presented here are preliminary. We are currently undertaking convergence study and detail study of the effects of cooling and star formation. In the near future, comparison of the high-resolution simulations of the kind presented here to high-resolution *Chandra* and *XMM-Newton* observations of merging clusters and cores of relaxed clusters should 1) constrain the relative importance of the physical processes and the dynamical processes operating during different stages of cluster evolution, 2) allow us to study metal enrichment of the ICM by supernova type Ia and II, and mixing of metals and their radial distribution, and 3) to shed new light on the puzzling absence of cold gas in the centers of classic "cooling flow" cluster.

D.N. acknowledges support from NASA GSRP Fellowship NGT8-52906. A.V.K. was partially supported by NSF grant No. AST-0206216 and by funding of the Center for Cosmological Physics at the University of Chicago (NSF PHY-0114422).

References

Durret, Lima Neto, G. B., Forman, W., and Churazov, E. (2003). An xmm-newton view of the extended "filament" near the cluster of galaxies abell 85. *accepted for publication in A&A(astro-ph/0303486).*

Ettori, S. and Fabian, A. C. (2000). Chandra constraints on the thermal conduction in the intracluster plasma of A2142. *MNRAS*, 317:L57–L59.

Forman, W., Jones, C., Markevitch, A., Vikhlinin, A., and Churazov, E. (2002). Galaxy clusters with chandra. *(astro-ph/0207165).*

Kravtsov, A. V. (1999). High-resolution simulations of structure formation in the universe. *Ph.D. Thesis.*

Kravtsov, A. V., Klypin, A., and Hoffman, Y. (2002). Constrained Simulations of the Real Universe. II. Observational Signatures of Intergalactic Gas in the Local Supercluster Region. *ApJ*, 571:563–575.

Markevitch, M., Ponman, T. J., Nulsen, P. E. J., Bautz, M. W., Burke, D. J., David, L. P., Davis, D., Donnelly, R. H., Forman, W. R., Jones, C., Kaastra, J., Kellogg, E., Kim, D. ., Kolodziejczak, J., Mazzotta, P., Pagliaro, A., Patel, S., Van Speybroeck, L., Vikhlinin, A., Vrtilek, J., Wise, M., and Zhao, P. (2000). Chandra observation of abell 2142: Survival of dense subcluster cores in a merger. *ApJ*, 541:542–549.

Markevitch, M., Vikhlinin, A., and Forman, W. R. (2002). A high resolution picture of the intracluster gas. *To appear in ASP Conference Series, (astro-ph/0208208).*

Nagai, D. and Kravtsov, A. V. (2003). Cold fronts in cdm clusters. *ApJ*, in press (astro-ph/0206469).

Nagai, D., Kravtsov, A. V., and Kosowsky, A. (2003). Effects of internal flows on sunyaev-zel'dovich measurements of cluster peculiar velocities. *ApJ*, in press (astro-ph/0208308).

Sun, M. and Murray, S. S. (2002). Chandra view of the dynamically young cluster of galaxies a1367 i. small-scale structures. *To appear in ApJ, Vol 576, 2002*, (astro-ph/0206255).

Vikhlinin, A., Markevitch, M., and Murray, S. S. (2001a). A Moving Cold Front in the Intergalactic Medium of A3667. *ApJ*, 551:160–171.

Vikhlinin, A., Markevitch, M., and Murray, S. S. (2001b). Chandra Estimate of the Magnetic Field Strength near the Cold Front in A3667. *ApJ*, 549:L47–L50.

Voit, G. M. and Bryan, G. L. (2001). Regulation of the X-ray luminosity of clusters of galaxies by cooling and supernova feedback. *Nature*, 414:425–427.

WHIM EMISSION AND THE CLUSTER SOFT EXCESS: A MODEL COMPARISON

Jonathan P.D. Mittaz[1], Richard Lieu[1] & Renyue Cen[2]

[1]*Department of Physics, UAH, Huntsville, AL35899*

[2]*Princeton University Observatory, Princeton University, Princeton, NJ 08544*

Abstract The confirmation of the cluster soft excess (CSE) by XMM-Newton has rekindled interest as to its origin. The recent detections of CSE emission at large cluster radii together with reports of OVII line emission associated with the CSE has led many authors to conjecture that the CSE is, in fact, a signature of the warm-hot intergalactic medium (WHIM). In this paper we test this scenario by comparing the observed properties of the CSE with predictions based on models of the WHIM. We find that emission from the WHIM is 3 to 4 orders of magnitude too faint to explain the CSE emission. The only possibility is the models if they are missing a large population of small density enhancements or galaxy groups, but this would place have severe ramifications on the baryon budget.

1. Introduction

The location of all the baryons existing at the current epoch is still somewhat of a mystery. Observationally, the total sum of baryons seen in stars, galaxies and clusters of galaxies ($\Omega_b = (2.1^{+2.0}_{-1.4})h_{70}^{-2}\%$, Fukugita, Hogan & Peebles 1998) is only about half of the number of baryons required by big bang nucleosynthesis models ($\Omega_b = (3.9 \pm 0.5)h_{70}^{-2}\%$ Burles & Tytler 1998) or from measurements of the cosmic microwave background ($\Omega_b = (4.4 \pm 0.4)h_{70}^{-2}\%$ Bennett et al. 2003). Recent cosmological hydrodynamical simulations have, however, shown that this missing 50% of baryons may be in the form of a warm ($10^5 - 10^7$K), tenuous medium (with overdensities between $\delta \sim 5 - 50$) existing in filaments formed during the process of large scale structure formation (e.g. Cen & Ostriker 1999). This medium is generally called the 'warm-hot intergalactic medium' or WHIM.

Ever since it was first proposed, the detection of the WHIM has been an important goal in astrophysics. To date there has been some success in finding WHIM like material. Far UV and soft X-ray absorption lines from the WHIM have been reported, though the majority of detections seem to be confined to

R. Lieu and J. Mittaz (eds.), Soft X-Ray Emission from Clusters of Galaxies and Related Phenomena, 171–183.

matter in our local group (e.g. Fang, Sembach & Canizares 2003, Nicastro et al. 2003, Mathur et al. 2003). At higher redshifts the situation is more controversial with a few detections having been reported. Possible emission from the the WHIM has also been observed, with some weak X-ray detections (e.g. Soltan, Freyberg & Hasinger 2002, Zappacosta et al. 2002) which are in approximate agreement with the predicted luminosity of the WHIM inferred from cosmological simulations. In general, however, these searches have not provided a strong, unambiguous detection of the WHIM either in absorption or emission at redshifts beyond our local group.

Recently, new observations of the cluster soft excess emission (CSE) show that this phenomenon may also be a signature of the WHIM. The CSE (seen as an excess of observed flux above the the hot intracluster medium (ICM) at energies below lkeV) has been an observational puzzle for a number of years. First discovered in the Virgo cluster (Lieu et al. 1996) subsequent observations have found similar behavior in a number of different systems (e.g. Lieu et al. 1996b, Mittaz et al. 1998, Kaastra et al. 1999, Bonamente et al. 2001a, Bonamente et al. 2001b, Bonamente et al. 2002). Observationally the CSE shows a number of different characteristics. In some clusters the CSE when expressed as a percentage is spatially constant (e.g. Coma: Lieu et al. 1998), in others there is a marked radial dependence with the fractional soft excess being stronger in the outer regions of the cluster (e.g. Mittaz et al. 1998). Observations have also shown that the CSE is a relatively common phenomena - a *ROSAT* study of 38 clusters showed that approximately 45% of clusters show at least a 1σ effect (Bonamente et al. 2002) .

On the current state of interpretation, models assuming a thermal origin of the excess have invoked a warm gas intermixed with the hot ICM are unsatisfactory. These have found to be unsatisfactory, however, since the cooling times are extremely short (sometimes $\sim 10^6$ years Mittaz et al. 1998) so some kind of heating mechanism to sustain the warm gas has to be envisaged (Fabian 1997). On the other hand if the emission arises from an inverse-Compton process (such as suggested by Hwang 1997, Sarazin & Lieu 1998, Ensslin, Lieu & Biermann 1998) then the radial dependence seen in some clusters can be explained, but the inferred pressure of the required cosmic-rays would be perplexedly too high (Lieu, Axford & Ip 1999).

The picture has changed somewhat with new observations by the XMM-Newton satellite. These observations reveal apparent strong thermal lines of Oxygen in the CSE spectra for the outskirts of a number of clusters (Kaastra et al. 2003, Finoguenov et al. 2003) with the CSE continuum being fitted with a characteristic temperature of 0.2 keV. If established, this provides irrefutable evidence that the CSE must be thermal in nature. Such a finding is also supported by *ROSAT* observations of the Coma cluster, which shows a very large, degree scale halo of soft emission (Bonamente et al. 2003). Emission on this scale cannot possibly be non-thermal in nature since there is no way of confining a relativistic particle population at such distances from the cluster center. Therefore, given that there seems to be a large scale thermal component at the outskirts of clusters, the CSE emitting material has to be located predominantly beyond the cluster virial radius where we can have sufficiently low densities to overcome any cooling time issues. In such a scenario the warm emitting gas is spatially consistent with the WHIM i.e. current cosmological models should be able to predict the luminosity and temperature of the CSE. We have therefore undertaken a detailed comparison between cosmological simulations of the WHIM and the observed soft excess signal.

2. The Model

We have used a recent cosmological hydrodynamic simulation of the canonical cosmological constant dominated cold dark matter model (Ostriker & Steinhardt 1995) with the following parameters: $\Omega_m = 0.3$, $\Omega_\Lambda = 0.7$, $\Omega_b\, h^2 = 0.017$, h = 0.67, $\sigma_8 = 0.9$ and a spectral index of the primordial mass power spectrum of n = 1.0. The simulation box has a size of 25 h^{-1} Mpc comoving on a uniform mesh with 768^3 cells and 384^3 dark matter particles giving a comoving cell size of 32.6 h^{-1} kpc. This simulation together with another similar one at lower resolution derived from the same code and parameters have previously been used to account for a variety of of different observational consequences of the WHIM, such as OVI absorption line studies (e.g. Cen, Tripp, Ostriker, Jenkins, 2001) and the X-ray background (Phillips, Ostriker & Cen, 2001). For a more detailed discussion of the simulation itself see Cen, Tripp, Ostriker & Jenkins (2001). Here we are primarily concerned with the emission in the vicinity of a cluster. Thus we focus our analysis on a cluster simulated in the high resolution mode.

The simulation as provided comes in the form of three data cubes containing temperature, density and metal abundance. Figure 1 shows an emission weighted temperature map from one particular projection of the cube and within the image a number of structures can be seen, including filaments, groups and one cluster candidate at a temperature of $\sim$ 6.5keV. For the purpose of this paper we are going to concentrate on this cluster-like structure which is seen in the top right corner of Figure 1. It has a peak mass density of 5.7×10^{13} $M_\odot$ Mpc^{-2} placing it the regime of virialized objects and so can be considered for our present purposes a cluster of galaxies.

3. Simulated X-ray spectra from the model

The ultimate aim of this paper is to investigate the WHIM model predicted X-ray emitting properties and compare them with current observations, particularly those of the CSE as now observed with XMM-Newton. To do this we

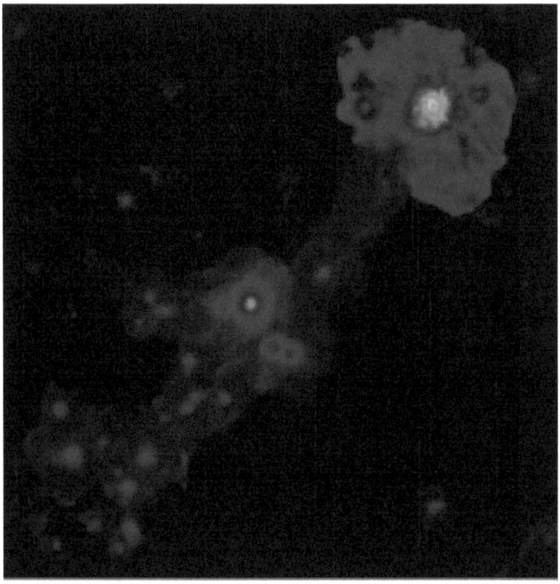

Figure 1. Emission weighted temperature map for one particular projection of the data volume. The plotted temperature range is from 0.1-6.5 keV. The simulated cluster can be seen in the top right of the figure

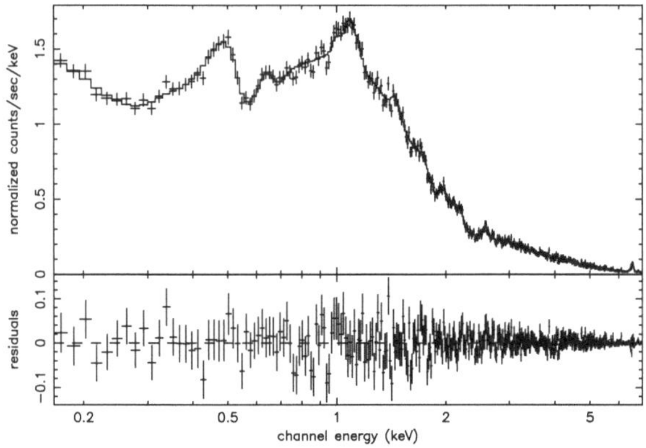

Figure 2. The simulated spectrum from the cluster showing the spectrum from the central 32.6 kpc region (0-2 arcminutes). Also shown is the best fit model (kT = 4.7 A=0.5) together with fit residuals and shows no requirement for any cluster soft excess

have placed the simulation at a redshift of 0.02, similar to that of the Coma cluster whose CSE has been extensively studied (Bonamente et al. 2003, Finoguenov et al. 2003). We further assumed that the gas is in thermal and

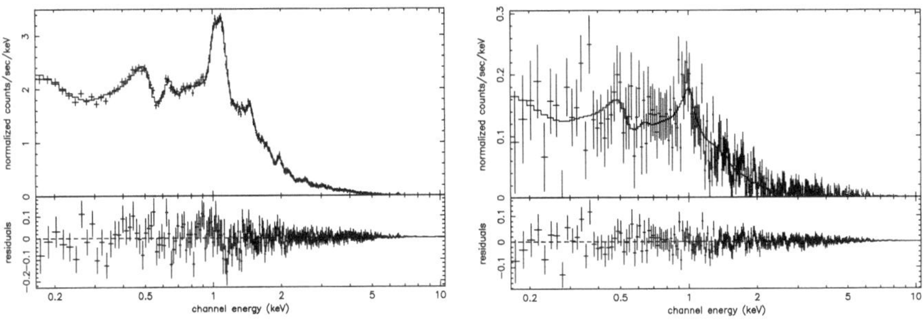

Figure 3. Further spectra from the simulated cluster from annular regions placed at the outskirts of the cluster. The left panel shows the 589-680 kpc annulus, and the right panel shows the simulated spectrum taken from the 1.17-1.26 Mpc annulus. Again, all spectra are well fitted with a single temperature model (kT = 2.27 keV A=0.29 and kT=1.25 A=0.06 respectively) with no evidence for any soft excess emission.

ionization equilibrium. Since the simulation provides the temperature, density and metal abundance at each point, for each cell we can derive an optically thin X-ray spectrum. To do this we have used the mekal code for optically thin plasmas (Kaastra et al. 1992, Leidhal et al. 1995) to generate the average bremsstrahlung and emission-line spectra. In order to avoid possible extreme abundance values that can sometimes be found in the densest regions of the simulation we do not allow the metal abundance to exceed a value of 0.5 solar. Then after adding the appropriate galactic absorption ($N_H = 9 \times 10^{19}$ cm^{-2}) and folding the resultant spectrum through the XMM-Newton MOS1 response we can generate an XMM-Newton counts spectrum from each cell. To obtain the overall spectrum we simply sum these spectra along our given line of sight. Finally, in order to get a reasonable estimate of the noise, an astrophysical background spectrum was added (taken from Lumb et al. 2002) together with Poisson fluctuations appropriate for the given exposure time (in our case 50 ksec).

Figure 2 shows an example of a simulation of the central 40 kpc (i.e. one cell which corresponds to 0-2 arcminutes) region of the cluster. Also shown is the best fit performance of the single temperature mekal emission model to the data with parameters of 4.7 keV and abundance of 0.5. As can be seen from the residual plot, there are no strong deviations of the data from the fit. This is consistent with general properties of clusters inferred from XMM-Newton and Chandra data (e.g. Molendi & Pizzolato 2001) where observationally there is no requirement for a multi-temperature fit, even within the so called cooling radius of the cluster where there is presumed to be a wide range of temperatures along the line of sight. As pointed out by Molendi & Pizzolato (2001) and

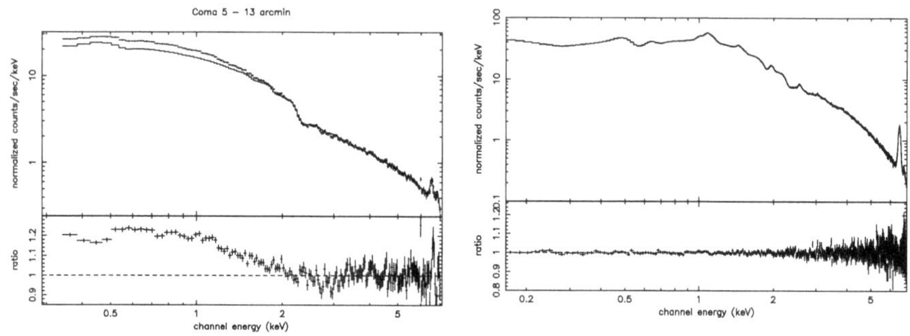

Figure 4. The left panel shows the XMM-Newton spectrum of Coma extracted in the 5-13 arcminute region (from Nevalainen et al. 2003), and the right panel shows a simulated spectrum from the model extracted from a similarly sized region. While the observed data shows a clear soft excess, the spectrum from the simulation show no need for any extra components above the hot ICM emission.

Ettori (2002) this is to be expected for a single phase medium which further supports our model choice.

For a more in depth understanding, the reason why we see such a good fit to the data even though there are multiple temperatures and densities along the line of sight, is due to the weighting effect of density. Since emission goes as n_e^2, the spectrum is heavily biased towards the densest material along the line of sight. Therefore the spectrum is dominated by the emission from a narrow range of densities and temperatures. Other components, such as may give rise to a CSE, will not be visible unless they exists in substantial quantities relating to the principle component.

3.1 Looking for the cluster soft excess

The sufficiency of a single temperature fit to the X-ray spectrum is seen at almost all radii ranging from the center to the outermost regions. This is illustrated in Figure 3 which shows examples from two outer annuli situated at 580-680 kpc and 1.17-1.26 Mpc from the center where the one temperature fits (with kT = 2.27 keV & kT = 1.25 keV respectively) are perfectly adequate. In the context of a WHIM explanation of the CSE this is interesting, because if the cluster soft excess is due to overlying filaments from the WHIM it is exactly in these outer regions (where the cluster emission is weak) that we would expect to see an effect. The fact the we see no evidence for a soft excess implies that the emission from the filaments in this simulation must be much weaker than that of the observed CSE.

The difference between observations and theory is even more clearly seen in figure 4 which shows two spectra: the spectrum on the left is taken from Nevalainen et al. (2002) and is of the 5-13 arcminute region of the Coma clus-

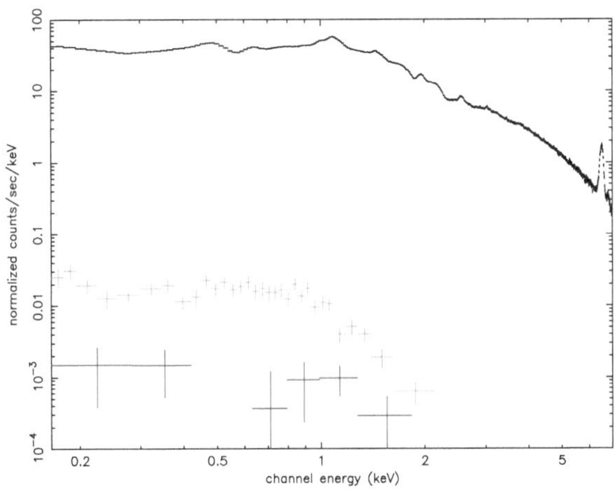

Figure 5. The total simulated spectrum from figure 4 (top curve) together with spectra derived from cells with temperatures below 1keV. The lowest spectrum is that derived from the same line of sight used for the total spectrum, while the middle spectrum is taken from a line of sight chosen to maximize the soft excess emission.

ter, the spectrum on the right is taken from the simulation covering a similar annular region. The first point to note is the relatively good agreement in the flux levels - the 0.2 - 2keV luminosity from the simulation is $\sim 4.5 \times 10^{44}$ ergs/s while the real observations shows a luminosity of $\sim 8.8 \times 10^{44}$ ergs/s, giving us confidence that the cross normalizations are realistic. However, there are also significant differences. Below 2 keV the XMM-Newton observation of Coma shows a clear soft excess at a 25% level above the hot ICM emission, whereas the simulated spectrum show no soft excess emission. We therefore conclude that the model does not include material at the right temperature and density to account for the thermal origin of the CSE.

Such a conclusion is further underpinned when the emissivity of material at a temperature consistent with the CSE is studied. Figure 5 shows the same simulated spectrum as the one in figure 4 but with the addition of spectra derived only from those cells with temperatures below 1 keV i.e. just those cells from which we would expect the cluster soft excess emission to arise. The two lower spectra in figure 5 show the expected emission from all components below 1keV as taken from two different lines of sight. The lower spectrum is derived from the same sight line used to create the total spectrum (i.e. the top spectrum). The brighter of the two spectra is taken from a sight line that has the most CSE effect, because along it the emission from cells with a temperature less the 1 keV is maximized i.e a line of sight that maximizes the CSE. Note in both cases there *is* emission in the crucial < 1 keV regime only it is much fainter (by a factors of 10^4 or 10^3 respectively) than the hot ICM emission. It then becomes obvious why no cluster soft excess was seen in the simulated spectra: the emission from cells capable of accounting for the soft excess is very small compared with the soft flux from the hot ICM emission lying behind it. We can further quantify this effect by comparing our model prediction with the soft excess emission seen in a large sample of clusters. Bonamente et al. (2002) studied 38 clusters and listed the strength of any soft excess detected with *ROSAT* . The simulated CSE luminosities fall short of the values for the majority of these cases.

3.2 A possible detection of a soft excess and the importance of small groups

Under certain circumstances it *is* possible to emulate a soft excess. For example from an extraction from the 30 - 45 cell (978 - 1467 kpc) annular region one notices a strong excess at the lowest energies. The left panel of Figure 6 shows the simulated spectrum, and in this case a single temperature fit is unacceptable with a χ^2_ν of 42.5. Following the analysis techniques of Nevalainen *et al.* (2002) and Bonamente *et al.* (2003) of applying a one temperature model to the spectrum above ~ 1 keV we find a reasonable fit with a single temperature of 1.08 ± 0.05 keV ($\chi^2_\nu = 1.28$). Below 1 keV there is then a very strong soft excess including a very strong OVII line. Interestingly the temperature of this excess is 0.2 keV, exactly the same temperature as has been reported for a number of clusters (eg. Kaastra *et al.* 2003). However, it turns out that this excess arises from a very small number of cells within the extraction region. In Figure 7 an image of the outskirts of the simulated cluster together with the annulus used to extract the original spectrum. Also shown is a small circle on the left hand side containing the region of bright pixels i.e. high densities, that actually gives rise to the soft excess. If this small region is removed, the resultant spectrum (the right hand panel of Figure 6) yields no evidence of a CSE.

We therefore conclude that there is very little evidence to link emission from the WHIM with the cluster soft excess. Our investigation has revealed the possibility, however, that small density enhancements (which may be associated with small galaxy groups) can give a signal that mimics the cluster soft excess. However, in the real observations the cluster soft excess seems relatively smooth and does not seem to be confined to a few small locations (see for example the *ROSAT* results on the Coma cluster Bonamente *et al.* 2003). One possibility is that the models to date do not have sufficient spatial resolution to emulate a large population of small density perturbations which could give rise to the observed soft excess signal. Additional physics related to a more complex environment, such as would exist in a supercluster, may also be required to achieve this - Kaastra et al. (2003) discussed the possibility, which clearly awaits further work. An overall fundamental problem exists, though,

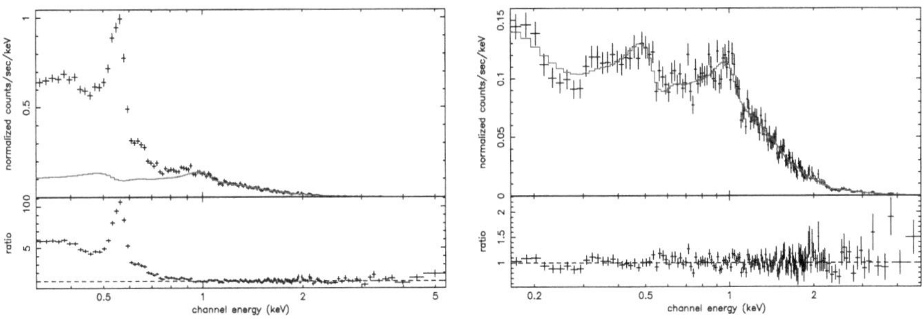

Figure 6. The left panel shows the simulated XMM-Newton spectrum for the annular region 978-1467 kpc showing a very strong soft excess signal. The right panel shows the same region if a small region containing a strong density enhancement is excluded. The fit now is acceptable over the entire XMM-Newton energy range ($chi^2_\nu = 1.18$) with a temperature of 1.07 keV

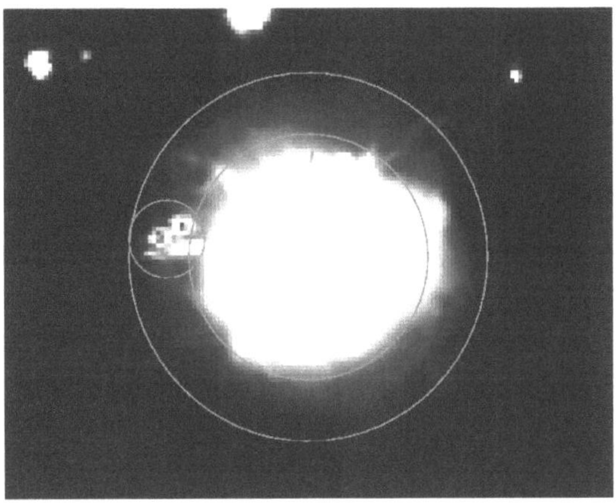

Figure 7. An image of the simulated cluster showing the 978-1467 kpc extraction annulus used to extract the spectrum from figure 6 together with the excluded region that contains all of the soft excess emission. The maximum count rate shown in the image is 0.01 cts/sec/pixel.

and it pertains to the acute lack of mass in gas at any temperature range to account for the soft excess. To get the required factor of 10^3 increase in brightness from the WHIM would imply an increase in the density by a factor of 30. Since the majority of the WHIM has overdensities between 10-30 we then require overdensities greater than 300. However, only $< 20\%$ of the WHIM bave overdensities greater than 200 (e.g. Davé et al. 2001) and it is clear that the models simply do not contain material at the densities and temperatures to give rise to the CSE signal as observed.

4. Other models

All of the above discussion has been based on the analysis of one particular model. It is reasonable to ask if other models show the same effect. This is actually quite a difficult question to answer since there are only a few models with all the required information freely available. A larger volume, lower resolution model that that studied in this paper is obtainable from Renyue Cen's web page (http://www.astro.princeton.edu/ cen/ PROJECTS/p1/p1.html) and we also utilized this version of the model. In general we find the same result - it is very difficult to reproduce the observed soft excess signal. With a larger volume there is, of course, a slightly higher probability of having galaxy groups located along any sight line which can give rise to excess soft flux in a similar fashion to the case studied in section 16.3.2. However, even by taking this effect into account we still arrive at a similar conclusions to those obtained form the higher resolution model - the WHIM itself cannot reproduce the observed CSE. We have also studied one of the clusters available in the Laboratory of Computational Astrophysics (LCA) simulated cluster archive (http://sca.ncsa.uiuc.edu) with no major change in the conclusions.

5. Discussion

From the above discussion it would appear that within our current understanding one cannot assign sufficient luminosity and low temperature material (< 1 keV) to account for an observable CSE using a WHIM filament. What are the possible reasons for this discrepancy? One possible problem with the high resolution simulation is that may not cover enough filaments/structures to emulate the CSE in the majority of cases. However, the discrepancy between model and observation is unlikely to be due simply to the scale of the simulation since the box side of 25 h^{-1} Mpc equals approximately that of the proposed filament needed to explain the Coma soft excess (Finoguenov et al. 2003). Therefore the scale of any filament in this simulation is not orders of magnitude discrepant from that required by observation. Further, a comparison of the temperatures and densities in cells in the vicinity of the cluster shows a distinct lack of cells with an overdensity > 200 (the value required by observations see e.g. Nevailenen et al 2002) and temperatures commensurate with that of the cluster soft excess ($T < 0.6$keV). It is not surprising then that the predicted soft emission is so weak - material at the required temperature and density is not present. Other simulations also seem to have the same absence of cells with the correct temperature/density ratio. In fact, the only time that a soft excess exists is during a strong density enhancement along the line of sight, which may be interpreted as a small galaxy group.

There are other possibilities that may point to this particular simulation as not representative of the clusters showing a CSE. Kaastra *et al.* (2003,2004) have proposed a correlation between the cluster soft excess and a supercluster environment. They arrive at this conclusion based on two arguments. *ROSAT* all-sky survey maps on degree scales around soft excess clusters seem to show a large scale excess of soft emission which is claimed to be related to a supercluster. The authors also claimed that in the model of Fang et al. (2002), regions where there are numerous structures (i.e. a potential simulated supercluster) can also be associated with a OVII column density similar to that inferred from observations ($\sim 4 - 9 \times 10^{16}$ cm^{-2}) thereby implying a causal link between superclusters and the presence of a CSE.

The first possibility, that of large scale soft emission, has already been demonstrated as being inconsistent with the models since the density of the WHIM and hence the emissivity is too low. We investigated the second argument using both the small and large scale simulations of Cen (scale 25 and 100 Mpc respectively) and have calculated the OVII column densities projected along the line of sight. In both cases we then compared the emission weighted temperature with the OVII column density to see where and how often the OVII is consistent with the strength of the line emission reported by Kaastra et al. The two models are shown in figure 8 and indicate that at locations where the projected emission weighted temperature is greater than 2 keV (which corresponds to locations where a cluster would exist) very few cells have the required OVII column. Indeed, for the smaller of the two models (the model with the higher resolution) there were no areas with an OVII column $> 4 \times 10^{16}$ cm^{-2} (the density required by Kaastra et al. 2003) and a temperature above 2 keV. For the larger, low resolution model 45 cells out of a total of 3568 consistent (or 1.26%) with a cluster had an OVII column large enough. While this may indicate that there is some relationship between a CSE the density of structures in the model, 1.26% is a very small fraction compared to the size and scale of the observed soft excess seen in real observations. From the point of view of absorption line studies, these plots also show that in the vicinity of clusters we would expect an OVI absorbing column of order 10^{14}cm^{-2} rather than the value predicted by Kaastra et al. (4×10^{16}cm^{-2}).

The only remaining option is that the XMM-Newton observations observe regions where there are a large number of small overdense, group like structures. However, in the current simulations a large number of such regions do not exist. We are then left with the following options. There could still be problems with the reality of the OVII detection by XMM-Newton and the excess is non-thermal. This, however, cannot alleviate the difficulty posed by the *ROSAT* Coma results of Bonamente et al. (2003). Therefore, there still may be something missing in the simulations, such as insufficient resolution. What seems clear is that a thermal interpretation of the CSE calls for more material,

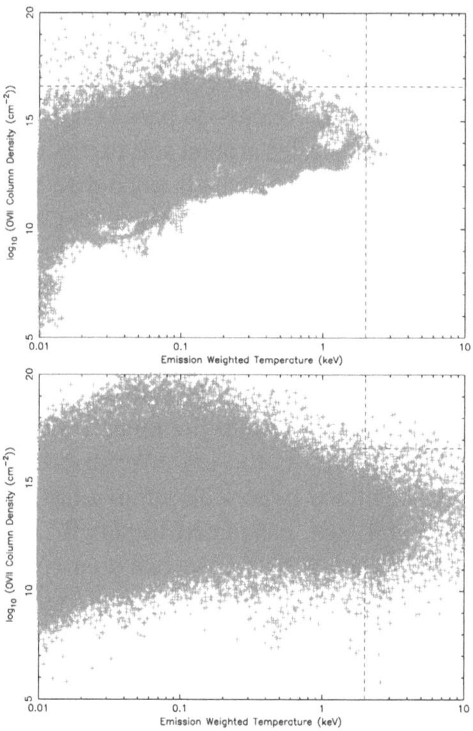

Figure 8. The left panel and right panels show the distribution of OVII column density as a function of the projected emission weighted temperature for both the small and large simulations respectively, In both plots the horizontal dotted line shows the observed lower limit to the OVII column density derived from XMM-Newton observations of the CSE. The vertical line shows the 2 keV point above which we have assumed that that particular location will correspond to a cluster. A CSE observation will correspond to point to the right and above the two lines but in both models there are far fewer cases (0% and 1.26% for the small and large model respectively) than would satisfy the observations

by a factor of at least 30, than is currently included in the models. If this is true, then the implication to the final overall budget for baryons is daunting

References

Bennett, C.L. et al., 2003, ApJS, 148, 1

Bonamente, M., Lieu, R. & Mittaz, J.P.D., 2001a, ApJLett, 552, 7

Bonamente, M., Lieu, R. & Mittaz, J.P.D., 2001b, ApJLett, 561, 63

Bonamente, M.,et al. 2002, ApJ, 576, 688

Bonamente, M., Lieu, R & Joy, M., 2003, ApJ, 595, 722

Burles, S. & Tytler, D., 1998, Space Sci. Rev., 84, 65

Cen, R. & Ostriker, J., 1999, ApJ, 514, 1

Cen, R., Tripp, T., Ostriker, J., Jenkins, E., 2001, ApJ, 559, 5C

Davé et al., 2001, ApJ, 552, 473
Ensslin, T.A, Lieu, R. & Biermann, P.L., 1999, A&A, 397, 409
Ettori, S.,et al., 2002, MNRAS, 330, 971
Fabian, A.C. 1997, Science, 275, 48
Fang, T., Bryan, G.L. & Canizares, C.R.. 2002, ApJ, 564, 604
Fang, T., Sembach, K.R. & Canizares, C.R., 2003, ApJLett, 586, 49
Finoguenov, A, Briel, U & Henry, P., 2003, A&A, 410, 777
Fukugita, M., Hogan, C.J.& Peebles, P.J.E, 1998, ApJ, 503, 518
Hwang, C, 1997, Science, 278, 1917
Kaastra, J.S. 1992, An X-Ray Spectral Code for Optically Thin Plasmas (Internal SRON-Leiden Report, updated version 2.0)
Kaastra, J. et al., 1999, ApJ, 519, 119
Kaastra, J., et al., 2003, A&A, 397, 445
Kaastra, J., et al., 2003b, astro-ph/0305424
Liedahl, D.A., Osterheld, A.L., & Goldstein, W.H., 1995, ApJL, 438, 115
Lieu, R., et al., 1996, ApJLett, 458, 5
Lieu, R., et al., 1996b, Science, 274, 1335
Lumb, D.,et al., 2002, A&A, 389, 93
Mathur, S., Weinberg, D. & Chen, X., 2003, ApJ, 582, 82
Mittaz, J.P.D, Lieu, R. & Lockman, J., 1998, ApJLett, 419, 17
Molendi, S. & Pizzolato, Fl, 2001, ApJ, 560, 194
Nevelienen, J.et al., 2003. ApJ, 584, 716
Nicastro, F., et al., 2003, Nature, 421, 719
Nicastro, F., 2003, astro-ph/0311162
Ostriker, J.P. & Steinhardt, P., 1995, Nature, 377, 600
Phillips, L.A., Ostriker, J.P. & Cen, R., 2001, ApJ, 554, 9
Sarazin, C. & Lieu, R., 1998, ApJLett, 494, 177
Soltan, A.M., Freyberg, M.J. & Hassinger, G., 2002, A&A, 395, 475
Zappacosta, L., et al., 2002, A&A, 394, 7

VI

NEW INSTRUMENTATION AND THE FUTURE

OBSERVING THE WARM-HOT INTERGALACTIC MEDIUM WITH XEUS

Jelle S. Kaastra[1] and Frits B.S. Paerels[2]

[1] *SRON National Institute for Space Research, Utrecht, The Netherlands*

[2] *Columbia University, New York, USA*

Abstract We discuss the potential for XEUS for observing the Warm-Hot Intergalactic Medium. After a short description of the mission we demonstrate the power of XEUS for detecting the WHIM both in emission and absorption.

1. Introduction

Soft excess X-ray emission in clusters of galaxies and related phenomena is the topic of this conference. In several contributions the role of non-thermal emission in clusters has been discussed. Here we focus upon the emission and absorption properties of the Warm-Hot Intergalactic Medium (WHIM) as it can be observed with XEUS.

XEUS is a project currently under study by ESA and Japan. A detailed account of several proposed aspects of the mission, both instrumental and scientific, can be found in the proceedings of a XEUS workshop (Hasinger et al. 2003, see http://wave.xray.mpe.mpg.de/conferences/xeus-workshop for an electronic version). A good overview of the potential for XEUS for studying the WHIM is given by Barcons (2003) and Paerels et al. (2003). In the present paper we present spectral simulations in order to demonstrate how XEUS can contribute to this field of research.

2. The XEUS mission

XEUS is a mission concept that is currently being studied by ESA and Japan. It can be viewed as a successor of XMM-Newton but with an order of magnitude better performance in several aspects. At the heart of the mission is a large X-ray telescope with a focal length of 50 m. Due to this length, the instruments are distributed over two satellites, one containing the mirror module and the other the detector module. Its limiting sensitivity is 200 times deeper than XMM-Newton. Following the launch of the initial spacecraft con-

R. Lieu and J. Mittaz (eds.), Soft X-Ray Emission from Clusters of Galaxies and Related Phenomena, 187–194.

figuration, an observation period of 4–6 years is planned. After this period, the satellites will have a rendez-vous with the International Space Station for refurbishment and addition of extra mirror area. Also the detector spacecraft can be replaced at that time. The satellite is then ready to observe for another period in this enhanced configuration.

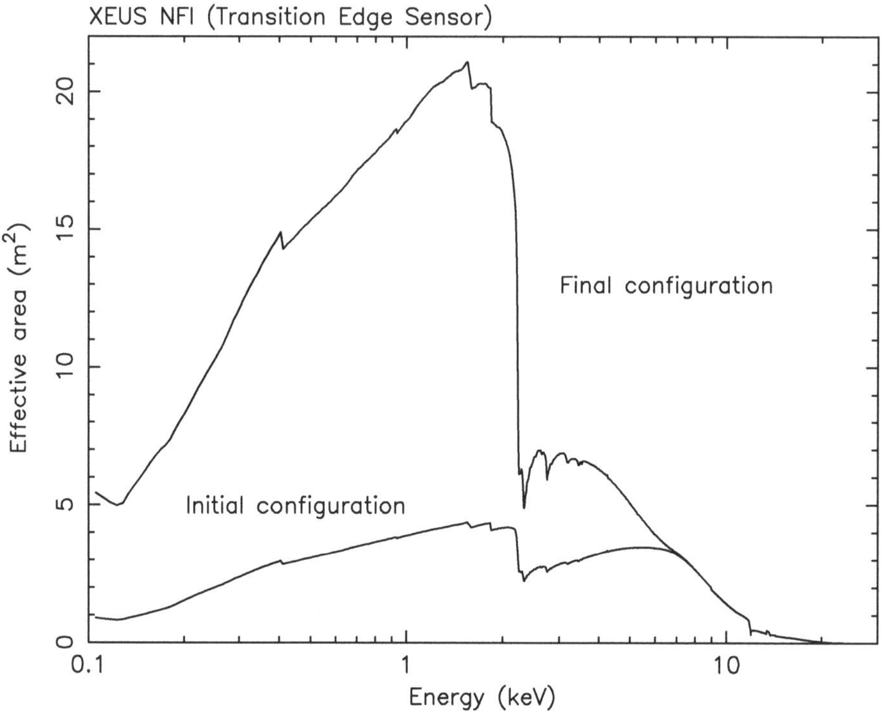

Figure 1. Effective area of XEUS (including mirror effective area, detector quantum efficiency and filter transmission, both for the initial (lower curve) and final (upper curve) configuration. The calculation is made for a Transition Edge Sensor.

The collecting area of the XEUS mirrors is very large, at 1 keV the area is 6 m^2 for the initial and 30 m^2 for the final configuration. The total effective area is shown in Fig. 1 for both configurations. Even in the initial configuration, the effective area is 20 times larger than the combined effective area of the three XMM-Newton EPIC camera's.

The imaging resolution is 2 arcsec Half Energy Width, with a field of view of 1 arcmin for the narrow field instruments (NFI) and 5 arcmin for the wide field instruments (WFI). For the narrow field instruments two types of detectors are being developed: Superconducting Tunnel Junctions (STJs) and Transition Edge Sensors (TESs). For the wide field instruments large area, small pixel size CCDs are envisaged.

The spectral resolution of these instruments varies between 1–5 eV for the NFI (between 1–8 keV) to 50–100 eV for the WFI over the same energy range. Currently studies are made in enhancing the sensitivity at higher energies even further. The nominal energy range is 0.05–30 keV.

In this paper we make simulations of emission and absorption spectra with XEUS. In all simulations, we have taken the final configuration with a TES detector, adopting a spectral resolution of 2 eV below 1 keV. The integration time for all simulations is taken to be 40 000 s.

3. The X-ray background

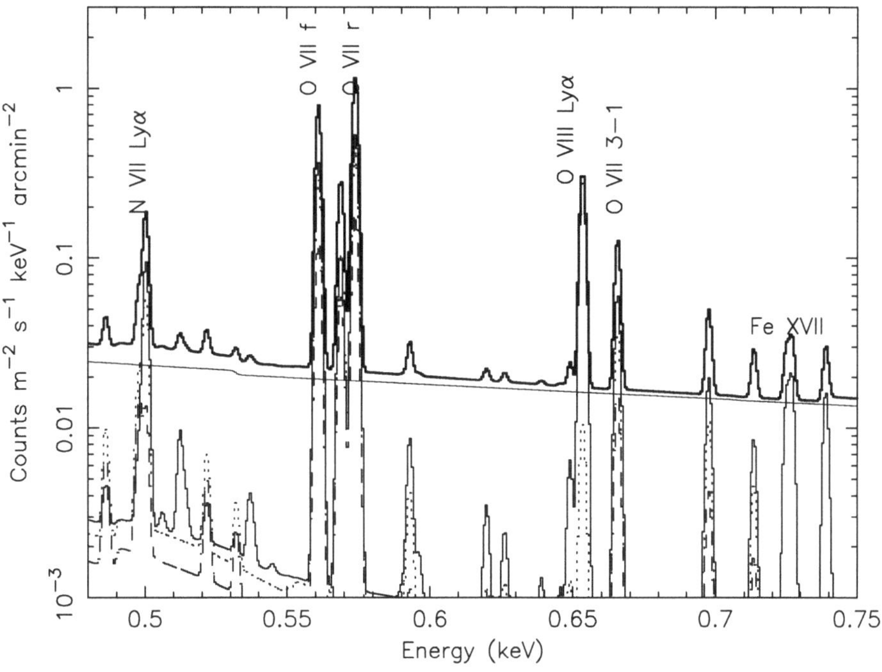

Figure 2. Portion of the predicted background spectrum for XEUS using the background model of Kuntz & Snowden (2000). Upper thick curve: total background; solid thin curve: extragalactic power law component; dotted line: Local Hot Bubble; dashed line: soft distant component; thin solid line with spectral lines: hard distant component.

In all spectra, but in particular in emission spectra of extended sources (due to a larger extraction radius), the Galactic foreground emission and extragalactic background emission must be taken into account in the effective "background" for the spectrum of the source that is studied. In Fig. 2 we show a portion of this simulated background spectrum, which represents a typical high Galactic latitude pointing. The model is based upon Kuntz & Snowden's (2000) decomposition of the soft X-ray background as estimated from Rosat

PSPC data. These authors decompose the background into four components: the Local Hot Bubble with a temperature of 0.11 keV; a soft distant component with a temperature of 0.09 keV; a hard distant component with a temperature of 0.18 keV; and the extragalactic power law with photon index 1.46. The last three components are affected by Galactic absorption. From Fig. 2 we see that while in the continuum most of the background around the important oxygen lines is caused by the extragalactic power law component, line emission from the other three components can be a factor of 10 stronger than the continuum at specific energies. In particular the hard distant component contributes a lot of flux in the background oxygen lines. At energies below 0.35 keV, most of the background flux is caused by the soft distant component and the Local Hot Bubble, the latter one dominating all other components below 0.2 keV. At these energies (below 0.2 keV) the spectral resolution of XEUS becomes poorer ($E/\Delta E < 100$) as compared to for example the LETGS of Chandra, but the effective area is orders of magnitude larger and detailed spectral analysis of the X-ray background (and any X-ray source) is still possible using the relative strengths of several line blends in a global fitting approach.

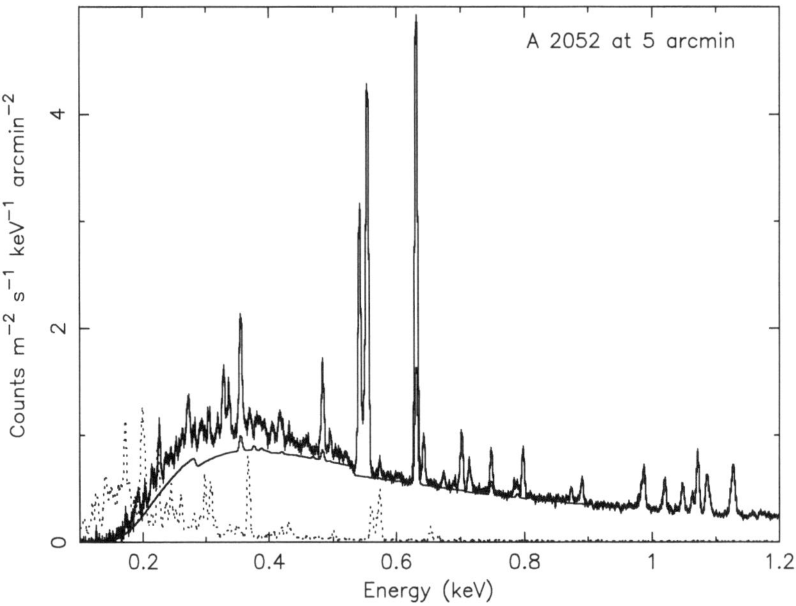

Figure 3. Simulated XEUS spectrum of the cluster of galaxies A 2052 (redshift 0.036) at 5 arcmin off-axis. The dashed line shows the subtracted background emission while the lower solid curve gives the contribution from the hot intracluster medium. The upper solid curve is the total (hot plus warm) background subtracted spectrum.

4. Emission spectra from the WHIM

With XMM-Newton the presence of thermal soft X-ray emission from the WHIM has been deduced from the presence of redshifted O VII line emission from a plasma with a temperature of 0.2 keV (Kaastra et al. 2003a; see also these proceedings: Kaastra et al. 2003b). Although the EPIC camera's of XMM-Newton have a much better spectral resolution than the PSPC detector used earlier in studying soft excess emission from clusters of galaxies, the resolution (60 eV) was still insufficient to resolve the O VII triplet into the resonance, intercombination and forbidden lines. In fact, the spectral resolution $E/\Delta E \sim 10$ at the oxygen lines only allowed a clear discrimination between a Galactic and extragalactic origin of the emission line by measuring its redshift, which was possible thanks to the sensitivity of XMM-Newton.

With XEUS and its high spectral resolution (2 eV) it is much easier to detect line emission and moreover, we can resolve the O VII triplet. This allows us to investigate the physical state of the warm gas in more detail, for example we can investigate the relative contribution from collisional ionization and photoionization by looking to the line ratio's of the triplet.

In Fig. 3 we show a simulated spectrum of the cluster A 2052 at 5 arcmin from the center of the cluster. The parameters for this simulation were taken from the best fit of Kaastra et al. (2003a). It is evident that it is very easy to detect the WHIM in or near such bright and nearby ($z = 0.036$) clusters, even taking into account that the simulation was done for a square box of 1×1 arcmin^2. Of course, the small field of view of the NFI makes it impossible to make detailed maps of such large clusters (radii of the order of 10–20 arcmin), but the enormous amount of information that is present in detailed investigations of selected pencil beams through the cluster compensate for this loss of field of view. Interestingly, with the XEUS sensitivity we do not only detect the O VII triplet from the WHIM, but are also sensitive to O VIII, Fe XVII as well as C VI and N VII line emission. In this way we can investigate the ionization structure as well as the chemical enrichment of the WHIM.

How far out can we do these kinds of studies? In order to simulate this, we have put A 2052 at a redshift of 1. The spectrum is now extracted over the entire annulus between 25–42 arcsec, which is smaller than the 1×1 arcmin^2 box of Fig. 3 but covers a relatively larger fraction of the cluster flux, due to the higher redshift. Note that the oxygen lines are now redshifted to $\sim$0.3 keV where the relative spectral resolution of XEUS is somewhat poorer; and due to the $1/r^2$ dependence of the flux, the cluster spectrum is now weaker than the subtracted background. Due to these effects, it is harder to disentangle the spectral components in the cluster spectrum. Detailed spectral fitting shows that we can detect the WHIM in emission with good confidence up to $z \sim 0.7$, and $z = 1$ is the real limit.

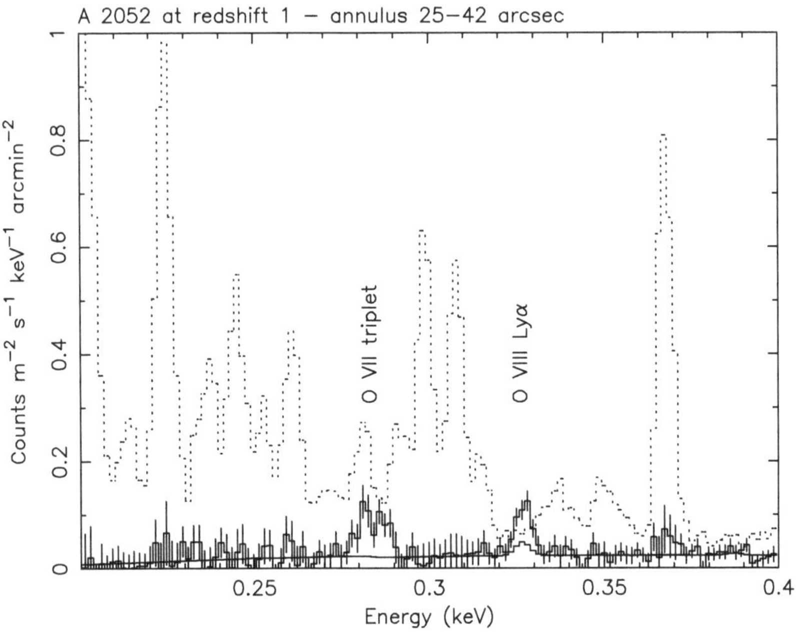

Figure 4. Simulated XEUS spectrum of the cluster of galaxies A 2052, but now put at a redshift of 1. The dashed line shows the subtracted background emission while the lower solid curve gives the contribution from the hot intracluster medium.

5. Observing the WHIM in absorption

In the previous section we focussed upon observing the WHIM in emission. Here we study the potential for XEUS for observing absorption lines. Simulations of the WHIM (for example Fang et al. 2002) show that at each random line of sight one expects to observe tens of absorbers with $z < 3$ and an O VIII column density larger than 10^{20} m^{-2}. Also a few absorbers with even higher column density ($> 10^{20.5}$ m^{-2}) are expected to occur along such lines of sight. So there are sufficient absorbers present. The question is if we have a sufficient number of bright background sources. Taking a flux limit of 10^{-16} W m^{-2} in the 0.5–4.5 keV band, we expect about 10 sources per square degree with this brightness. Hence this is a very common flux level and at almost any region of the sky that is of interest, XEUS will have at least one such source relatively nearby to be used as a background lamp. These sources have on average a redshift of 0.5–1.0, and according to the numerical simulations mentioned above they should therefore have several absorbers in the line of sight. We made a simulation of such a source with a photon index of 2 and absorbed by a typical Galactic column density (2×10^{24} m^{-2}). The result is shown in Fig. 5. It is evident that XEUS can easily detect O VIII columns of $10^{20.5}$ m^{-2} in these

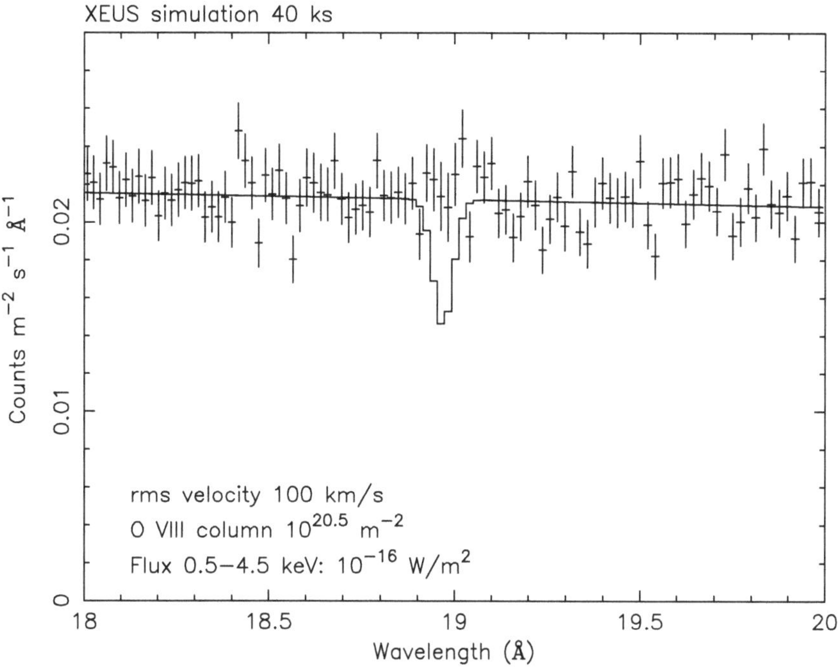

Figure 5. Simulated X-ray spectrum for a pure power law with a 0.5–4.5 keV flux of 10^{-16} W m^{-2} and Galactic absorption. The solid line indicates the model spectrum, but now absorbed by a slab of pure O VIII with a column density of $10^{20.5}$ m^{-2}.

relatively weak sources in reasonable integration times. For smaller column densities (10^{20} m^{-2}) larger exposure times than the current 40 ks are needed.

In brighter sources XEUS is even more sensitive. Fig. 6 shows a similar simulation as Fig. 5, but now for a source that is 100 times brighter. There are tens of these objects on the sky to study. In these objects, O VIII column densities of 10^{20} m^{-2} are easily detected and in fact the detection limit is at around 10^{19} m^{-2} in 40 ks. Hence, for these lines of sight it is possible to obtain detailed spectral information on the physical parameters of all relevant absorption systems. This is evident for example from the fact that we also see higher order lines such as Lyβ.

6. Conclusions

XEUS is the most sensitive instrument proposed thus far that can detect the WHIM, both in emission and absorption. Due to the fact that the emission scales with the square of the density, of course the densest regions are most easy detected. Emission studies are mostly hampered by the narrow field of view (1 arcmin). XEUS has a very high sensitivity for absorption lines, it can detect columns of 10^{20} m^{-2} of O VIII in relatively weak sources. With XEUS

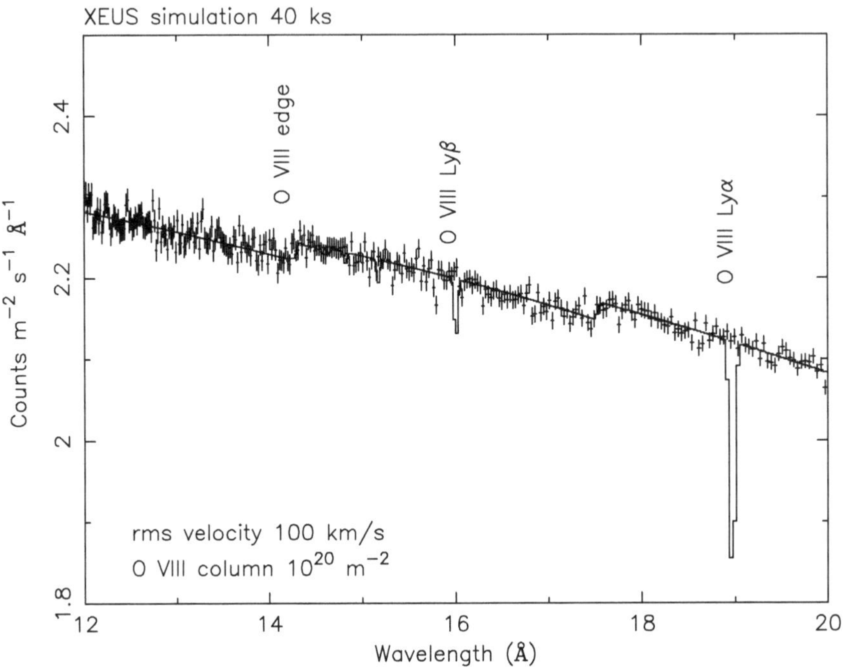

Figure 6. Simulated X-ray spectrum for a pure power law with a 0.5–4.5 keV flux of 10^{-14} W m^{-2} and Galactic absorption. The solid line indicates the model spectrum, but now absorbed by an additional slab of pure O VIII with a column density of 10^{20} m^{-2}.

it will be possible to obtain deep snapshots of the WHIM, both in emission and absorption.

Acknowledgements SRON is supported financially by NWO, the Netherlands Organization for Scientific Research.

References

Barcons, X., 2003, MPE Report 281, p. 77

Fang, T., Bryan, G.L., & Canizares, C.R., 2002, ApJ, 564, 604

Hasinger, G., Boller, T., & Parmar, A.N., 2003, MPE Report 281

Kaastra, J.S., Lieu, R., Tamura, T., Paerels, F. B. S., & den Herder, J. W., 2003a, A&A, 397, 445

Kaastra, J.S., Lieu, R., Tamura, T., Paerels, F. B. S., & den Herder, J. W., 2003b, these proceedings

Kuntz, K.D., & Snowden, S.L. 2000, ApJ, 543, 195

Paerels, F., Rasmussen, A., Kahn, S.M., den Herder, J.W., & de Vries, C., 2003, MPE Report 281, p. 57

Astrophysics and Space Science Library

Volume 302: ***Stellar Collapse,*** edited by Chris L. Fryer
Hardbound, ISBN 1-4020-1992-0, April 2004

Volume 301: ***Multiwavelength Cosmology,*** edited by Manolis Plionis
Hardbound, ISBN 1-4020-1971-8, March 2004

Volume 300: ***Scientific Detectors for Astronomy,*** edited by Paola Amico, James W. Beletic, Jenna E. Beletic
Hardbound, ISBN 1-4020-1788-X, February 2004

Volume 299: ***Open Issues in Local Star Fomation,*** edited by Jacques Lépine, Jane Gregorio-Hetem
Hardbound, ISBN 1-4020-1755-3, December 2003

Volume 298: ***Stellar Astrophysics - A Tribute to Helmut A. Abt,*** edited by K.S. Cheng, Kam Ching Leung, T.P. Li
Hardbound, ISBN 1-4020-1683-2, November 2003

Volume 297: ***Radiation Hazard in Space,*** by Leonty I. Miroshnichenko
Hardbound, ISBN 1-4020-1538-0, September 2003

Volume 296: ***Organizations and Strategies in Astronomy, volume 4,*** edited by André Heck
Hardbound, ISBN 1-4020-1526-7, October 2003

Volume 295: ***Integrable Problems of Celestial Mechanics in Spaces of Constant Curvature,*** by T.G. Vozmischeva
Hardbound, ISBN 1-4020-1521-6, October 2003

Volume 294: ***An Introduction to Plasma Astrophysics and Magnetohydrodynamics,*** by Marcel Goossens
Hardbound, ISBN 1-4020-1429-5, August 2003
Paperback, ISBN 1-4020-1433-3, August 2003

Volume 293: ***Physics of the Solar System,*** by Bruno Bertotti, Paolo Farinella, David Vokrouhlický
Hardbound, ISBN 1-4020-1428-7, August 2003
Paperback, ISBN 1-4020-1509-7, August 2003

Volume 292: ***Whatever Shines Should Be Observed,*** by Susan M.P. McKenna-Lawlor
Hardbound, ISBN 1-4020-1424-4, September 2003

Volume 291: ***Dynamical Systems and Cosmology***, by Alan Coley
Hardbound, ISBN 1-4020-1403-1, November 2003

Volume 290: ***Astronomy Communication,*** edited by André Heck, Claus Madsen
Hardbound, ISBN 1-4020-1345-0, July 2003

Volume 287/8/9: ***The Future of Small Telescopes in the New Millennium,*** edited by Terry D. Oswalt
Hardbound Set only of 3 volumes, ISBN 1-4020-0951-8, July 2003

Volume 286: ***Searching the Heavens and the Earth: The History of Jesuit Observatories,*** by Agustín Udías
Hardbound, ISBN 1-4020-1189-X, October 2003

Volume 285: ***Information Handling in Astronomy - Historical Vistas,*** edited by André Heck
Hardbound, ISBN 1-4020-1178-4, March 2003

Volume 284: ***Light Pollution: The Global View,*** edited by Hugo E. Schwarz
Hardbound, ISBN 1-4020-1174-1, April 2003

Volume 283: ***Mass-Losing Pulsating Stars and Their Circumstellar Matter,*** edited by Y. Nakada, M. Honma, M. Seki
Hardbound, ISBN 1-4020-1162-8, March 2003

Volume 282: ***Radio Recombination Lines,*** by M.A. Gordon, R.L. Sorochenko
Hardbound, ISBN 1-4020-1016-8, November 2002

Volume 281: ***The IGM/Galaxy Connection,*** edited by Jessica L. Rosenberg, Mary E. Putman
Hardbound, ISBN 1-4020-1289-6, April 2003

Volume 280: ***Organizations and Strategies in Astronomy III,*** edited by André Heck
Hardbound, ISBN 1-4020-0812-0, September 2002

Volume 279: ***Plasma Astrophysics , Second Edition,*** by Arnold O. Benz
Hardbound, ISBN 1-4020-0695-0, July 2002

Volume 278: ***Exploring the Secrets of the Aurora,*** by Syun-Ichi Akasofu
Hardbound, ISBN 1-4020-0685-3, August 2002

Volume 277: ***The Sun and Space Weather,*** by Arnold Hanslmeier
Hardbound, ISBN 1-4020-0684-5, July 2002

Volume 276: ***Modern Theoretical and Observational Cosmology,*** edited by Manolis Plionis, Spiros Cotsakis
Hardbound, ISBN 1-4020-0808-2, September 2002

Volume 275: ***History of Oriental Astronomy,*** edited by S.M. Razaullah Ansari
Hardbound, ISBN 1-4020-0657-8, December 2002

Volume 274: ***New Quests in Stellar Astrophysics: The Link Between Stars and Cosmology,*** edited by Miguel Chávez, Alessandro Bressan, Alberto Buzzoni,Divakara Mayya
Hardbound, ISBN 1-4020-0644-6, June 2002

Volume 273: ***Lunar Gravimetry,*** by Rune Floberghagen
Hardbound, ISBN 1-4020-0544-X, May 2002

Volume 272:***Merging Processes in Galaxy Clusters,*** edited by L. Feretti, I.M. Gioia, G. Giovannini
Hardbound, ISBN 1-4020-0531-8, May 2002

Volume 271: ***Astronomy-inspired Atomic and Molecular Physics,*** by A.R.P. Rau
Hardbound, ISBN 1-4020-0467-2, March 2002

Volume 270: ***Dayside and Polar Cap Aurora,*** by Per Even Sandholt, Herbert C. Carlson, Alv Egeland
Hardbound, ISBN 1-4020-0447-8, July 2002

Volume 269: ***Mechanics of Turbulence of Multicomponent Gases,*** by Mikhail Ya. Marov, Aleksander V. Kolesnichenko
Hardbound, ISBN 1-4020-0103-7, December 2001

Volume 268: ***Multielement System Design in Astronomy and Radio Science,*** by Lazarus E. Kopilovich, Leonid G. Sodin
Hardbound, ISBN 1-4020-0069-3, November 2001

Volume 267: ***The Nature of Unidentified Galactic High-Energy Gamma-Ray Sources,*** edited by Alberto Carramiñana, Olaf Reimer, David J. Thompson
Hardbound, ISBN 1-4020-0010-3, October 2001

Volume 266: ***Organizations and Strategies in Astronomy II,*** edited by André Heck
Hardbound, ISBN 0-7923-7172-0, October 2001

Volume 265: ***Post-AGB Objects as a Phase of Stellar Evolution,*** edited by R. Szczerba, S.K. Górny
Hardbound, ISBN 0-7923-7145-3, July 2001

Volume 264: ***The Influence of Binaries on Stellar Population Studies,*** edited by Dany Vanbeveren
Hardbound, ISBN 0-7923-7104-6, July 2001

Volume 262: ***Whistler Phenomena - Short Impulse Propagation,*** by Csaba Ferencz, Orsolya E. Ferencz, Dániel Hamar, János Lichtenberger
Hardbound, ISBN 0-7923-6995-5, June 2001

Volume 261: ***Collisional Processes in the Solar System,*** edited by Mikhail Ya. Marov, Hans Rickman
Hardbound, ISBN 0-7923-6946-7, May 2001

Volume 260: ***Solar Cosmic Rays,*** by Leonty I. Miroshnichenko
Hardbound, ISBN 0-7923-6928-9, May 2001

Volume 259: ***The Dynamic Sun,*** edited by Arnold Hanslmeier, Mauro Messerotti, Astrid Veronig
Hardbound, ISBN 0-7923-6915-7, May 2001

Volume 258: ***Electrohydrodynamics in Dusty and Dirty Plasmas- Gravito-Electrodynamics and EHD,*** by Hiroshi Kikuchi
Hardbound, ISBN 0-7923-6822-3, June 2001

Volume 257: ***Stellar Pulsation - Nonlinear Studies,*** edited by Mine Takeuti, Dimitar D. Sasselov
Hardbound, ISBN 0-7923-6818-5, March 2001

Volume 256: ***Organizations and Strategies in Astronomy,*** edited by André Heck
Hardbound, ISBN 0-7923-6671-9, November 2000

Volume 255: ***The Evolution of the Milky Way- Stars versus Clusters,*** edited by Francesca Matteucci, Franco Giovannelli
Hardbound, ISBN 0-7923-6679-4, January 2001

Volume 254: ***Stellar Astrophysics***, edited by K.S. Cheng, Hoi Fung Chau, Kwing Lam Chan, Kam Ching Leung
Hardbound, ISBN 0-7923-6659-X, November 2000

Volume 253: ***The Chemical Evolution of the Galaxy***, by Francesca Matteucci
Paperback, ISBN 1-4020-1652-2, October 2003
Hardbound, ISBN 0-7923-6552-6, June 2001

Volume 252: ***Optical Detectors for Astronomy II***, edited by Paola Amico, James W. Beletic
Hardbound, ISBN 0-7923-6536-4, December 2000

Volume 251: ***Cosmic Plasma Physics***, by Boris V. Somov
Hardbound, ISBN 0-7923-6512-7, September 2000

Volume 250: ***Information Handling in Astronomy***, edited by André Heck
Hardbound, ISBN 0-7923-6494-5, October 2000

Volume 249: ***The Neutral Upper Atmosphere***, by S.N. Ghosh
Hardbound, ISBN 0-7923-6434-1, July 2002

Volume 247: ***Large Scale Structure Formation***, edited by Reza Mansouri, Robert Brandenberger
Hardbound, ISBN 0-7923-6411-2, August 2000

Volume 246: ***The Legacy of J.C. Kapteyn***, edited by Piet C. van der Kruit, Klaas van Berkel
Paperback, ISBN 1-4020-0374-9, November 2001
Hardbound, ISBN 0-7923-6393-0, August 2000

Volume 245: ***Waves in Dusty Space Plasmas,*** by Frank Verheest
Paperback, ISBN 1-4020-0373-0, November 2001
Hardbound, ISBN 0-7923-6232-2, April 2000

Volume 244: ***The Universe***, edited by Naresh Dadhich, Ajit Kembhavi
Hardbound, ISBN 0-7923-6210-1, August 2000

Volume 243: ***Solar Polarization***, edited by K.N. Nagendra, Jan Olof Stenflo
Hardbound, ISBN 0-7923-5814-7, July 1999

Volume 242: ***Cosmic Perspectives in Space Physics***, by Sukumar Biswas
Hardbound, ISBN 0-7923-5813-9, June 2000

Volume 241: ***Millimeter-Wave Astronomy: Molecular Chemistry & Physics in Space,*** edited by W.F. Wall, Alberto Carramiñana, Luis Carrasco, P.F. Goldsmith
Hardbound, ISBN 0-7923-5581-4, May 1999

Volume 240: ***Numerical Astrophysics,*** edited by Shoken M. Miyama, Kohji Tomisaka,Tomoyuki Hanawa
Hardbound, ISBN 0-7923-5566-0, March 1999

Volume 239: ***Motions in the Solar Atmosphere,*** edited by Arnold Hanslmeier, Mauro Messerotti
Hardbound, ISBN 0-7923-5507-5, February 1999

Volume 238: ***Substorms-4,*** edited by S. Kokubun, Y. Kamide
Hardbound, ISBN 0-7923-5465-6, March 1999

Volume 237: ***Post-Hipparcos Cosmic Candles,*** edited by André Heck, Filippina Caputo
Hardbound, ISBN 0-7923-5348-X, December 1998

Volume 236: ***Laboratory Astrophysics and Space Research,*** edited by P. Ehrenfreund, C. Krafft, H. Kochan, Valerio Pirronello
Hardbound, ISBN 0-7923-5338-2, December 1998

Missing volume numbers have not yet been published.
For further information about this book series we refer you to the following web site:
http://www.wkap.nl/prod/s/ASSL

To contact the Publishing Editor for new book proposals:
Dr. Harry (J.J.) Blom: harry.blom@wkap.nl

MIX
Papier aus verantwortungsvollen Quellen
Paper from responsible sources
FSC® C105338

If you have any concerns about our products,
you can contact us on
ProductSafety@springernature.com

In case Publisher is established outside the EU,
the EU authorized representative is:
Springer Nature Customer Service Center GmbH
Europaplatz 3, 69115 Heidelberg, Germany

Printed by Libri Plureos GmbH
in Hamburg, Germany